Advanced Technologies for Fluid-Particle Systems

Hamid Arastoopour, *Volume Editor*

Ye Mon Chen, Dimitri Gidaspow, Leon Glicksman Desmond King, and Wen-Ching Yang, *Volume Co-Editors*

Akilli, Huseyin
Alvarez, Mario M.
Arastoopour, Hamid
Bernstein, Barry
Berrino, Attilio
Bilgili, Ecevit
Bolland, Olav
Buettner, Helen M.
Buyevich, Y. A.
Chang, S. L.
Choi, Kyu Yong
Cody, G.D.
de Lasa, H.
Dave, Rajesh N.
Dudukovic, M. P.
Dunphy, William
Golchert, B.
Gu, Weikai
Hasan, Omar S.
Hatano, Hiroyuki
Herbert, Peter
Hjertager, Bjørn H.
Hoffman, Alex C.
Horio, Masayuki
Kamiya, Hidehiro
Kaneko, Yasunobu
Kapbasov, S.K.
Kim, Ju Yong
Krol, S.
Leckner, Bo
Lettieri, Paola
Levy, Edward K.
Mathiesen, Vidar
Matsuda, Satoru
Muzzio, Fernando J.
Newton, David
Nicolai, Rainer
Petrick, M.
Pfeffer, Robert
QI, Dewei
Ray, Madhumita Bhowmick
Reh, Lothar
Shiojima, Takeo
Skouby, David C.
Solberg, Tron
Tsuchiya, Katsumi
Watano, Satoru
Yates, John G.

AIChE Staff
Maura N. Mullen, Publications Production Manager
James L. Ivey, Publications Production Assistant
Armand Veneziano, Cover Design

AIChE Symposium Series

Number 321 1999 Volume 95

Published by

American Institute of Chemical Engineers

3 Park Avenue New York, N.Y. 10016-5901

AIChE shall not be responsible for statements or opinions advanced in their papers or printed in their publications.

Library of Congress CIP No. 99-047988

ISBN 0-8169-0815-X

FOREWORD

This volume of the AIChE Symposium Series continues the tradition of an annual volume presenting recent research and development in fluidization and fluid-particle systems. The papers were selected by peer reviews from those presented at the Advanced Technologies for Fluid-Particle Systems Symposium of the topical conference on Advanced Technologies for Particle Processing at the AIChE Annual Meeting in Miami Beach, Florida, from November 15 through 20, 1998.

This year's symposium volume contains papers that cover a wide range of topics related to fluid-particle systems and fluidization including:

- Fundamentals of Fluidization and Fluid-Particle Systems
- Mathematical Modeling and Computer Simulation of Fluid-Particle Systems
- Advances in Fluid/Particle Flow Property Measurement
- Applications of Particle Technology in Polymer and Rubber Processing
- Circulating Fluidized Beds
- Particle Interaction and Mixing

I would like to acknowledge all the chairs and co-chairs of the sessions for their careful selection of papers for the Miami Beach meeting; Dr. Mike Roco for his commitment and for an excellent job in organizing the topical conference; and my co-organizers of the Advanced Technologies for Fluid-Particle Systems Symposium: Drs. Yutaka Tsuji and Paul Nelson. I would also like to thank Maura N. Mullen and James L. Ivey of the AIChE publications staff, and the fluidization and particle technology research teams at Illinois Institute of Technology, whose critical review and further evaluation of the papers ensured the high quality of the papers in this volume. Finally, I would like to express my thanks to my assistant, Simone Bradford, without whose diligence, quality control and hard work, this volume would not have been possible.

Hamid Arastoopour, *Volume Editor*
Professor and Chairman
Department of Chemical and Environmental Engineering
Illinois Institute of Technology
10 W. 33rd Street
Chicago, Illinois 60616

CONTENTS

Foreword iii

Computational Analysis of Some Fluidized Systems
Bjørn H. Hjertager, Vidar Mathiesen, and Tron Solberg 1

Particle Fluctuation Velocity in Gas Fluidized Beds – Fundamental Models Compared to Recent Experimental Data
G.D. Cody, S. K. Kapbasov, and Y.A. Buyevich 7

On the Shape of Grade Efficiency Curves Computed for Centrifugal Dedusters by CFD with Lagrangian Particle Tracking
Alex C. Hoffman, Attilio Berrino, and Madhumita Bhowmick Ray 13

Investigation of Particle/Gas Flow Characteristics in Long Pipelines
B. Golchert, S. L. Chang, and M. Petrick 18

Saltation Phenomena in the Inlet Regions of Horizontal Pipes
Huseyin Akilli and Edward K. Levy 24

Multiphase Opaque Systems Experiments and Simulations
M.P. Dudukovic 30

Simulations of Sedimentation of Cylindrical Multiparticles in a Three-Dimensional Space
Dewei Qi 36

Diameters of Catalyst Clusters in FCC
Weikai Gu 42

Particle Clustering in Down Flow Reactors: Role of the Hydrodynamic Particle Interaction
S. Krol and H. de Lasa 48

Describing Mass Transfer in Circulating Fluidized Beds by Ozone Decomposition
Olav Bolland and Rainer Nicolai 52

The ETH Experience: Experimental Database and Results from the Past 8 Years
Peter Herbert, Rainer Nicolai, and Lothar Reh 61

Hydrodynamic Studies in a 0.45-m Riser with Liquid Feed Injection
David C. Skouby 67

DEM Simulation of Fluidized Beds of Polyolefin Synthesis
Yasunobu Kaneko, Takeo Shiojima, Hidehiro Kamiya, and Masayuki Horio 71

Polymer Particle Mixing and Segregation in a Gas Phase Olefin Polymerization Reactor
Ju Yong Kim and Kyu Yong Choi 77

Solid State Shear Extrusion (SSSE) Pulverization of Rubber Granules
Ecevit Bilgili, Hamid Arastoopour, and Barry Bernstein 83

Characterization of Suspension Flows Using Particle Image Velocimetry (PIV)
Omar S. Hasan, Mario M. Alvarez, Fernando J. Muzzio, and Helen M. Buettner 90

Continuous Measurement of Velocity and Direction of Particles Using the Lag-Time Method
Hiroyuki Hatano, Satoru Matsuda, and Katsumi Tsuchiya 95

On the Fluidization Behaviour of Some Industrial Materials at High Temperature: The Effect of Interparticle Forces
Paola Lettieri, David Newton, and John G. Yates 100

Fluidized Bed Combustion – Areas of Development
Bo Leckner 106

Dry Particle Coating by a Newly Developed Rotating Fluidized Bed Coater
Satoru Watano, Robert Pfeffer, Rajesh N. Dave, and William Dunphy 112

Index 117

Computational Analysis of Some Fluidized Systems

Bjorn H. Hjertager[2] Vidar Mathiesen[2,3], Tron Solberg[1,2]
[1]Aalborg University Esbjerg (AAUE), Niels Bohrs Vej 8, DK-6700 Esbjerg, Denmark
[2]Telemark Technological R&D Centre (Tel-Tek), Kjolnes Ring, N-3914 Porsgrunn, Norway
[3]Telemark College (HiT-TF), Kjolnes Ring, N-3914 Porsgrunn, Norway

Progress in modelling and simulation of flow processes in gas/particle systems carried out at the author's research groups are presented. Particular emphasis is given to computational fluid dynamics (CFD) models that use the multi-dimensional multi-fluid techniques. Turbulence modelling strategies for gas/particle flows based on the kinetic theory for granular flows are given. Submodels for the interfacial transfer processes and chemical kinetics modelling are presented. An overview of the numerical solution method used is also given. Examples are shown for several gas/particle systems including flow in risers, segregation by size and reacting systems.

A single screw extruder was used to pulverize vulcanized rubber granules in the solid state into fine particles under high shear and normal forces. This process, known as Solid State Shear Extrusion (SSSF), is non-cryogenic and has the capability of producing rubber particles having diameters ranging from 40 microns to about 1 500 microns. Because the rubber granules were partially de-vulcanized under high shear forces the produced particles were very suitable for recycling. During pulverization, a significant part of the strain energy which has been stored in the rubber at high shear and normal forces is released in the form of heat. This causes the temperature of the produced powder to increase. Rubber particles at high temperature and under normal forces can initiate an oxidation process and formation of undesired agglomerates. At a very high temperature, the particles may even degrade. This means the rate of heat transfer is one of the most critical parameters in controlling the sizes of the produced particles. Thus, this study focuses not only on the fundamentals of pulverization and processing of the particles but also on experimental and numerical analyses of heat transfer and the material temperature distribution throughout the process.

INTRODUCTION

The new multi-dimensional computer techniques for fluid flow processes (CFD) have made it possible to develop new design methods based on mathematical modelling of the basic processes that occur in the actual geometrical model of the reactor. To apply these methods the basic conservation equations of flow, chemical reactions and heat transfer as well as the coupling between them and the coupling between the phases must be solved for a multi-dimensional calculation domain. Some review papers dealing with CFD analysis in chemical engineering have appeared. The papers by Hjertager [1], Kuipers and van Swaiij [2], and Kuipers et al [3] give overviews of applications related to chemical reaction engineering. Several books dealing with gas/particle flows and fluidization have also appeared: Gidaspow [4], Fan and Zhu [5] and Crowe et al.[6].

Gidaspow [7] has reviewed the CFD models for gas/particle reactors that had evolved up until about 1985. The models at that time had demonstrated the capabilities to simulate some of the crucial characteristics like the bubble formation and development in bubbling fluidized beds. The models assumed a two-phase gas-particle flow with given apparent viscosities in the gas and in the particle phases. Shih et al. [8] introduced a model that take account of the multi-sized particle problem and showed application to a sedimentation problem. Mathiesen et al [9, 10, 11] have used a multi size model to analyse segregation in various risers based on an extension of the kinetic theory of granular flow to multi-sized particles (Gidaspow et al [12], Manger [13]). The work of Tsuo and Gidaspow [14] showed that the model was able to calculate the flow patterns in circulating fluidized beds. This model had to assume an apparent viscosity for the particle phase to be able predict the experimentally observed phenomena. The works of Ding and Gidaspow [15] and Ma and Ahmadi [16, 17] propose to couple the apparent viscosities in the two phases by introducing turbulence models for each of the phases. The works of Samuelsberg et al [18, 19] have continued this work. Kuipers et al [20] have analysed the flow of gas/particles in two-dimensional fluidized bed. A comprehensive model for two phase flows, heat transfer and chemical reactions in a fluidized bed gasifier has been presented by Gidaspow et al. [21]. Manger et al [13, 22, 23, 24] have formulated the multi-phase flow equations in curvilinear coordinates enabling analyses of complex geometries.

A variety of models have been proposed in order to design catalytic commercial-scale fluidized bed reactors from experimental findings in small scale units. Relatively few studies have been reported which uses multi-phase CFD models to simulate chemical reactions in catalytic fluidized bed reactors. Gidaspow et. al [25] studied the IGT-U-GAS coal gasification process and predicted many of the features of fluidized bed gasifiers. Theologos and Markatos [25, 26] used a three dimensional model to simulate a catalytic cracking riser type reactor. With a simplified kinetic model they simulated among other variables the yields and temperature distribution inside the reactor. Gidaspow and Therdthianwong [27] studied the effect of gas velocity of SO_2 sorption with calcined limestone in an isothermal CFB loop. Samuelsberg and Hjertager [28, 29, 30] simulated an isothermal ozone decomposition reaction in a circulating fluidized bed reactor. The predictions were compared against available ozone decomposition data. The predicted radial and axial ozone concentration profiles fitted the experimental data well, in both shape and magnitude. Samuelsberg and Hjertager [31] have analysed an exothermal oxy-chlorination reaction in a large scale bubbling bed reactor.

The aim is to present multidimensional prediction/design models for the chemical reaction and transport processes in multi-phase flows gas/particle chemical reactors. Finally, some examples of flow and chemical reaction modelling of fluidized bed reactors done by the authors and coworkers will be given.

GOVERNING EQUATIONS

The numerical model of the turbulent two-phase flow is based on the mult-fluid model (Spalding [32], Hjertager [33]). Transport equations governing the conservation of mass, momentum and scalar variables like turbulent quantities (θ_s for gas-particle flows), enthalpy (h_k) and concentration of chemical species ($Y_{j,k}$) is derived for each phase. The multi-fluid conservation equations governing the mean turbulent flow field for mass, momemtum and scalars are written as follows:

$$\frac{\partial}{\partial t}(\alpha\rho)_k + \nabla \cdot (\alpha\rho\vec{V})_k = \Gamma_k \tag{1}$$

$$\frac{\partial}{\partial t}(\alpha\rho\vec{V})_k + \nabla \cdot (\alpha\rho\vec{V}\,\vec{V})_k = \nabla \cdot (\alpha\underline{\sigma})_k - \alpha_k \nabla p + \vec{F}_k + \alpha_k\rho_k g \tag{2}$$

$$\frac{\partial}{\partial t}(\alpha\rho\Phi)_k + \nabla \cdot (\alpha\rho\vec{V}\Phi)_k = -\nabla \cdot (\alpha\vec{J}_\Phi)_k + S_{\Phi,k} \tag{3}$$

Here the subscript k refers to phase number k. k is either gas (g) or particle size number, solids (s). The pressure, p is considered to be common for the various phases. Turbulent diffusive transport of momentum and scalar variables are denoted $\underline{\sigma}_k$ and $\mathbf{J}_{\Phi,k}$, respectively. Interphase momentum transfer is denoted F_k and interphase mass transfer into phase no. k is denoted Γ_k.

Single particle phase The shear stress in the gas phase is related to the gradients of velocity components.The turbulent viscosity in the gas phase is modelled with use of a Sub Grid Scale (SGS) model proposed by Deardorff [34]. The conservation equations for the solids phase are based on kinetic theory for granular flow based on the work of Jenkins and Savage [43], Lun et al.[36] and further extended by Ding and Gidaspow [15] and Gidaspow [4]. This model makes it possible to calculate a granular temperature and, based on this granular temperature, a solids phase shear viscosity is explicitly computed. The shear stress in the solids phase is the sum of a collisional part and a kinetic part. The solids phase pressure which includes both kinetic and collisional pressures, is determined from an equation of state similar to the van der Waals equation of state for gases Chapman and Cowling [35]. In this expression the coefficient of restitution, *e,* is introduced along with the radial distribution function, g_0 (Lun et al [36]). In these expressions Θ is the granular temperature related to the kinetic turbulent energy of the particle motion. The solids phase shear viscosity for dense and dilute flow is a product of the mean free path times an oscillation velocity times a density.The granular temperature or turbulent kinetic energy Θ, is determined by solving a conservation equation of the form given in Equation (3). Here terms on the right side of the equation represent gradient diffusive transport, production due to shear, disssipation due to inelastic collisions and dissipation due to fluid friction.

Multiple particle phases Gidaspow et al [12] and Manger [13] have extended the kinetic theory for granular flow to multiple particle phases. This means that particles of different diameters or different densities are allocated separate phases. Each phase (particle phase) has separate sets of velocities, granular temperatures, etc. The coupling between the various particle phases is through particle pressures, radial distribution function, viscosities, particle collision dissipation and conductivity. The single particle expressions are extended to include sums over the total number of solids phases, n_s. In these sums the mass of a particle is the "weighting function". Full account of the multiple particle phase kinetic theory for granular flow is given by Manger [13], Mathiesen et al [9,10] and Mathiesen [11].

Interfacial forces For the gas (g) and solids (s) momentum equations the interphase friction is based on the Ergun equation for gas volume fraction $\alpha_g \leq 0.8$ and based on empirical correlation for $\alpha_g > 0.8$. For more than one particle phase we must in addition to drag between the gas phase and the particle phase, take account of particle-particle drag. This drag force is a rather complex expression and may be found in Manger [13], Mathiesen et al [9, 10] and Mathiesen [11].

Interfacial heat transfer Heat may be generated by the catalytic exothermic solids phase chemical reaction. This heat is transported between the phases and may in addition be cooled by a submerged heat exchanger. Appropriate source term for the gas phase and the solids phase are taken from Gidaspow [4]. The volumetric heat transfer coefficient is calculated by using several different correlations for the Nusselt number in different flow regimes, characterized by the Reynolds number and the gas phase volume fraction. Typical relationships for the bed to wall heat transfer coefficient may be found in Geldart [37].

Chemical reaction kinetics modelling For an ozone decomposition reaction the chemical reaction may be described as a simple homogeneous one step first order catalytic reaction. With the very low concentrations of ozone involved in the experiments of Ouyang et al. [38, 39], the reaction heat produced during the reaction is negligible and the reaction can be looked upon as isothermal. For an oxy-chlorination reaction the energy equations for the two phases are solved together with equations for the chemical species. The catalytic reactions take place on the surface of the catalyst. A simple one step homogeneous catalytic oxy-chlorination reaction can be represented by a first order reaction with the specific reaction rate given by the usual Arrhenius expression.

NUMERICAL PROCEDURE

Solution of the set of partial differential equations given above is performed by finite-domain methods. The calculation method is taken from the work of Spalding [32] in the form given by Hjertager [33]. The calculation domain is devided into a finite number of main grid points where pressure, densities, void fractions, turbulent quantities, enthalpies and mass fractions of the chemical species are stored. The velocity components are on the other hand stored at grid points located midway between the main grid nodes. The relevant conservation equations are integrated over control volumes surrounding the relevant grid points in space and over a time interval. This integration is performed using upwind differencing in space and implicit differencing in time. The result of this is a set of non-linear algebraic equations which are solved by application of the well known Tri-Diagonal Matrix Algorthm (TDMA) used along the various coordinate directions. For the void fraction and velocity equations a point iteration method is used. Special care has been taken to solve the coupling of the momentum equations with the continuity equations. For this we use the socalled IPSA procedure proposed by Spalding [32]. The Partial Elimination Algorithm (PEA) is used to decouple the drag force between the phases in the momentum equations. The algorithm and models given above are incorporated into a computer code named FLOTRACS-MP-3D. The software package consists of a preprocessing part, the actual computer code and a postprocessing part. The preprocessing part is based on a text editor or on AutoCAD (FloCAD-3D). The preprocessing part generates a case file, which gives input to the model and flow simulation. In the case file, input parameters are set including: geometry description, coordinate system, number of control volumes, number of phases, initial and boundary conditions as well as other flow parameters. The core of the software is the transient three-dimensional computer code, which uses the described mathematical model and solution algorithms. For the postprocessing part various commercially available software packages are used. Some of these include UNIGRAPH, UNIMAP, AGX, EXPLORER.

PREDICTION OF FLOW CHARACTERISTICS IN GAS/ PARTICLE FLUIDIZED BEDS

The model for gas/particle flow given above has been verified in several isothermal circulating fluidized beds by Hjertager and coworkers. A short review of this work will be given here and more detail may be found in the cited references.

Mean and turbulence velocity Samuelsberg and Hjertager [18] have measured and predicted the mean and turbulence velocity profiles in a circulating bed of 30mm in diameter, 1m in height. They used Laser Doppler anemometry to measure the solids flow velocities. It was found that the predicted particle velocities have the same general trend as measured. The particles moved upwards in the core and downwards near the walls. A behaviour which is typical for core-annulus flow. For gas superficial velocity $V_{g,sup} = 0.71$ m/s it was shown how the velocity profile compares well with the experimental results in the core of the reactor at all heights. The core annulus flow in the simulations gave the general trend, but a little too high central velocity at the highest level. The downflow near the wall was over predicted at all heights. The simulations were not quite symmetrical at the two lowest heights. This may be due to the asymmetrical backflow of particles. There was a slight acceleration of the particles throughout the riser, which indicates that the flow was not fully developed. The RMS velocity was derived from the granular temperature through the relation $V_{RMS} = (3\times\Theta)^{1/2}$. A comparison between the simulated RMS velocity and the measured RMS velocity was therefore possible. The general trend in the simulations was in relatively good agreement with the experimental results, although the simulations were a little too low at all levels. The simulated profiles were not quite symmetrical for any of the simulations. For the case with superficial gas velocity of 0.71 m/s, it was found that experimental RMS velocity had the steepest gradients at 0.16 m, and more evenly distributed at the higher levels. This general trend was also given by the simulations, except that the velocities were a little too low.

Solids distribution and behaviour Samuelsberg and Hjertager [19] have also predicted the mean velocity and volume fraction profiles in a circulating bed of 75mm in diameter, 7m in height. The experimental data was provided by Miller and Gidaspow [40]. The simulated average volume fractions equaled the experimental, which is 0.03. Although the inlet boundary conditions were steady, the flow never reached a steady state. The gas and solids would oscillate with a frequency that varied with gas superficial velocity and solids flux. These oscillations increased with increasing riser height and increasing feed flux and would reach a maximum at the top of the riser due to less dampening effect in the upper part. The slugs that could be observed visually were also predicted in the simulations. The mean flux of these slugs could grow to more than three times the inlet mass flux.

Recent progress in turbulence and multiphase modelling Manger [13] and Mathiesen et al [9, 10] have extended the model of Samuelsberg and Hjertager [29, 30, 31] to include more particle phases and an improved multi particle kinetic theory. The multi-particle size model was generalized to analysis of one gas phase and n_s number of solids phases to enable description of realistic particle size distributions. Each solids phase was characterized by a diameter, form factor, density and restitution coefficient. The granular temperature and momentum equations were solved for each phase. Mathiesen et al [9, 10] did 2D simulations using one gas phase and three solids phases. The circulating fluidized bed was 2 m high and had a cross section of 0.2 x 0.2 m. The experimental data were given by Azario et al [41]. The simulations were done with different superficial gas velocities, initial solids concentration and standard deviations of the particle size distributions. Here emphasis will be given to the results dealing with the effects of different particle size distributions and to study the fluctuating behaviour of the gas/solids flow system. The CFD model was shown to capture large scale fluctuations of the mean velocities, volume fractions and mean diameter. Axial and radial large scale fluctuations were calculated and added to the small scale turbulence obtained from the isotropic granular temperature. It was demonstrated that the computed small scale turbulence was to small compared to the measurements. However, by adding the large scale fluctuations to the small scale turbulence a correct order of magnitude was obtained. It was also shown that the calculations confirm that the axial fluctuations were much larger than the radial ones. Simulations were performed with two different particle size distributions and compared against experimental data along the centre line of the riser. The results showed that the computed mean diameters were in good agreement with experimental data both in form and magnitude. For the narrow size distribution the particle diameter was almost constant along the centre line, whereas for the wide-range particle size distribution a significant segregation by size was correctly obtained.

Species concentration distribution Samuelsberg and Hjertager [29] have predicted the isothermal ozone decomposition in a circulating bed of 250mm in diameter, 11m in height. The experimental data was taken from Ouyang et al [38, 39]. Radial profiles of measured and simulated dimensionless ozone concentrations at 4 different axial locations were used for comparing experiments and simluations. Both measurements and simulations gave the same trend, the ozone concentration was higher in the centre than close to the reactor walls. This was consistent with the observed solids volume fractions profiles, since the conversion is directly proportional to the solids volume fraction. At the highest axial position, the profile was almost flat, the difference in ozone concentration between core and wall region decreases with increasing height. The simulated results fitted the experimental results well.

Prediction of complex and 3D geometries

Manger et al [22, 23] has extended the model of Samuelsberg and Hjertager [18, 19] to include geometry description based on curvilinear coordinates. Mathiesen [11] has extended the model to include three-dimensional geometry description.

The ticking hourglass Manger et al [22] used a two-dimensional model with a curvilinear description to calculate the ticking hour glass. Experimental data are given by Wu et al [42]. The simulations showed qualitative agreement with experimental findings. Like the measurements the simulations indicate an oscillatory flow of sand through the orifice of the hour glass with a well defined period of the oscillations. Both experiments and simulations indicate that the oscillatory flow is induced by a gas bubble which is formed in the solids phase at the orifice. This could be seen in the sequence of predicted void fraction contours. The bubble rises through the sand and breaks at the surface. The simulations and experimental data of total mass through the orifice show qualitatively similar oscillatory flow. The flux through the orifice is an intermittent flow with a well-defined period of oscillations. It is found that for $D/d_p > 49$ ($d_p < 41$ μm), the oscillations decay and eventually the flow through the orifice becomes steady.

Three-dimensional flow simulation in a riser Mathiesen [11] has extended the model to include three-dimensional geometry description. The experimental setup by Azario et al [41] previously analysed by a 2D approximation given above, was analysed by a full 3D geometry description. It should be noted that no experimental verification of the predictions was tried. Since the actual riser section of the circulating fluidized bed was rectangular, a three dimensional Cartesian geometry description gave the best representation of the riser. The grid was uniform in the axial direction, whereas a non-uniform grid was used for the two transverse directions. In Figure 1 a short time series of the total volume fractions of solids is presented. The horizontal as well as the vertical plot are made after 6, 8 and 10 seconds. The horizontal planes are plotted 0.3, 1.0 and 1.6 m above the inlet and the vertical plot is for y=0.1 m. The volume fraction distributions is chaotic and random at all heights, but most clear in the dense turbulent bed. The dense region seems to be somewhat higher than 1.0 m. There are relatively large concentration gradient which are related to the bubble formation in the riser. Inside the bubbles there is relative low solids fraction, whereas the solids concentration is very high in the emulsion phase. Small bubbles are formed in the corners as well as in the rest of the cross-section area, particular at the lowest height. The time averaged turbulent velocities for each solids phase have also been predicted. The turbulent velocity was derived from the granular temperature and did not include contributions from the large scale fluctuations. An interesting observation in the simulation was that the largest particles have a considerably higher turbulence velocity than the smaller and intermediate ones. These phenomena are explained by the fact that gas turbulence will be lower than the particle fluctuations and that smaller particles will follow the gas flow more easily than larger particles.

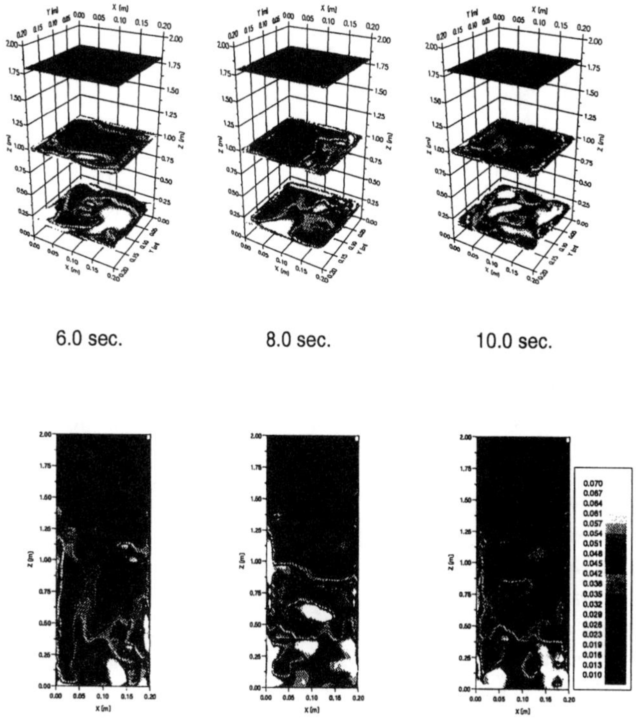

Figure 1. Time series of the total volume fraction of solids.

Prediction of oxychlorination reaction in a gas particle fluidized bed

A simplified oxy-chlorination reaction was simulated by Samuelsberg and Hjertager [31] in a full-scale bubbling bed reactor to demonstrate the model's capabilities and possibilities to simulate gas/solids flows with exothermic chemical reaction and heat transfer between the phases as well as to internal heat transfer equipment. The reactor was 4 metres in diameter and 20 metres high. It was simulated using a non-uniform Cartesian mesh with 42x68 gridnodes in the radial and axial direction respectively. Initially the reactor was filled with an 8-metre high bed of FCC catalysts with a particle diameter of $60 \cdot 10^{-6}$ m and density of 1600 kg/m^3. The inlet gas superficial velocity was held constant at 0.4 m/s, at the outlet boundary a continuation condition was used for the gas phase. No particles were allowed to leave the reactor. The inlet mass fractions of the chemical species participating in the reaction were the same as the initial mass fraction in the reactor. The initial temperature for the gas and solids present in the reactor was the same as the inlet gas temperature which equaled 200°C. The reactor gauge pressure was set to 5 bars. The reaction heat in this strongly exothermic reaction was removed by generation of steam in submerged heat exchangers. The internal heat transfer equipment, which will be hundreds of submerged tubes in a commercial reactor, was modelled as a porous block with a volume blockage of 0.75, this was approximately corresponding to 200-250 pipes with a diameter of 0.15 m. The temperature in the porous block was held constant at 200°C, and both phases were able to exchange heat with the heat exchange device.

The integrated flow, reaction and heat transfer model was capable of simulating the formation and propagation of gas bubbles in a full scale bubbling bed reactor. The gas and solids circulation pattern set up by these bubbles influence the gas-solids temperature and the chemical species distribution in the bed. The exothermic reaction model was capable to predict the formation and destruction of the chemical species involved in the chemical reaction. An overall steady reaction state was gained after 120 seconds of real simulation. A conversion based on the limiting reactant of 0.974 is achieved, which gave an outlet mass fraction of EDC of 0.47. The largest part of the chemical reaction was shown to occur in the first metres above the inlet. The large amount of heat generated in this reaction was effectively cooled with the aid of a submerged heat exchange device, modelled as a porous block. The time averaged temperature in the bed ranged from 250°C just above the inlet to a temperature of 230°C at the outlet boundary. Due to the very large heat transfer coefficient between the phases involved, the temperature difference between the phases is negligible. The zones close to the wall were not participating in the reaction and are only used in axial transport of cooler gas and particles from the upper part of the reactor to the lower part. This transport increased the temperature gradients and improved the cooling of the reactor walls, even though the reactor volume must be larger than needed for the reaction only, due to this recirculation. Already produced EDC recirculated unnecessary by the axial transport back to the high temperature zone which may accelerate by-product formation.

CONCLUDING REMARKS

A comprehensive multidimensional CFD model for turbulent gas particle flow, heat transfer and chemical reaction in fluidized bed reactors was presented. Due to the discrepancies found between the experimental and the simulated results, and to be able to simulate commercial processes for reliable scale-up or optimization purpose, further improvements of the model are needed. The recommendations for further work can be summarized as follows: The flow model should be extended to three dimensions using curvilinear coordinates to make the model able to capture three-dimensional effects in order to obtain more detailed information of the processes involved. The model for granular temperature should be extended to include the influence of a multi-sized particle mixture on solids rheology and to account for particle packing. The chemical reaction model should be extended in order to account for heterogeneous and/or side-reactions in a fluidized bed reactor. The model for mass and heat transfer should be further developed and tested against available data. In order to obtain sharp interfaces, higher order discretizations scheme should be incorporated. Further testing and verification of the combined flow, chemical reaction and heat transfer model against reliable experimental data is vital for the use and acceptance of such models in both research and industrial practice.

ACKNOWLEDGEMENTS

The work on fluidized bed reactor modelling at Tel-Tek is sponsored by the Norwegian Research Council (NFR), Norsk Hydro and Statoil. The authors are grateful to Dr. K. Morud, Dr. A. Samuelsberg, Dr. E. Manger and Dr. P.C. Friberg former PhD students as well as Mr. T. Solbakken and Ms. T. Teppen, PhD students at Telemark Institute of Technology that all have contributed to the work.

REFERENCES

[1] Hjertager, B.H. (1997), MIC - Modelling, Identification and Control, Vol 18, No 1, pp 19-59.

[2] Kuipers, J.A.M, van Swaiij, W.P.M (1998), "CFD applied to chemical reaction engineering", Adv. in Chem. Eng.

[3] Kuipers, J.A.M, Hoomans, B.P.B., van Swaiij, W.P.M (1998), in Fluidization IX, pp 15-30.

[4] Gidaspow, D (1994), Multiphase flow and fluidization: Continuum and kinetic theory descriptions, Academic press.

[5] Fan, L.S, Zhu, C. (1998) Principles of gas-solids flows, Cambridge Univ. Press.

[6] Crowe C., Sommerfeldt, M., Tsjui, Y. (1998) Multiphase flows with droplets and particles, CRC Press.

[7] Gidaspow, D (1986), Appl. Mech. Rev., Vol 39, No. 1, pp. 1-23.

[8] Shih, Y.T., Gidaspow, D and Wasan, D.T. (1987), Powder Technology, Vol. 50, pp. 201-215.

[9] Mathiesen, V., T. Solberg, E. Manger, B.H. Hjertager (1996), 5th Int. Conf. on CFB, May 28-31, Beijing, China.

[10] Mathiesen, V., T. Solberg, B.H. Hjertager (1998), Proc. of 3rd Int. Conf. on Multiphase Flow, Lyon, France, June 8-12.

[11] Mathiesen, V. (1997), Dr. Ing. Thesis, Telemark Institute of Technology, Porsgrunn, Norway

[12] Gidaspow, D., Huilin, L., Manger, E. (1996), XIXth Int. congr. of theor. and appl. mech., Kyoto, Japan, Aug. 25-31, 1996.

[13] Manger, E. (1996), Dr.Ing. Thesis, Telemark Institute of Technology, Porsgrunn, Norway.

[14] Tsuo, Y.P and Gidaspow, D. (1990), AIChE Journal, Vol. 36, pp. 885-896.

[15] Ding, J. and Gidaspow, D. (1990), AIChE Journal, Vol. 36, pp. 523-538, 1990.

[16] Ma, D., Ahmadi, G. (1990), Int. journal of multiphase flows, vol.16, no.2, pp. 323-340.

[17] Ma, D. and Ahmadi, G. (1990), Int. J. Multiphase Flows, Vol. 16, pp. 342-351.

[18] Samuelsberg, A., B.H. Hjertager (1996), Int. J. Multiphase flow, Vol. 22, No. 3, pp. 575-591.

[19] Samuelsberg, A., B.H. Hjertager (1996), AIChE Journal, Vol. 42, No. 6, pp. 1536-1546.

[20] Kuipers, J.A.M., Van Duin, K.J., van Beckum, F.P.H., van Swaaij, W.P.M. (1993), Comp. Chem. Eng., Vol 17, pp 839-899.

[21] Gidaspow, D., Ettehadieh, B. and Bouillard, J. (1985), AIChE Symp. Series No. 241, Vol. 80, pp. 57-64.

[22] Manger, E., T. Solberg, B.H. Hjertager, D. Vareide (1995), Int. J. Multiphase Flow, Vol. 21, No. 4, pp. 561-567.

[23] Manger, E., T. Solberg , B.H. Hjertager (1995), Proc. of 5th Int. Conf. on Bulk Materials Handling and Transportation, Newcastle, N.S.W., Australia, 9-12 July, Vol. 1, pp. 199-204, 1995.

[24] Manger, E., T. Solberg, B.H. Hjertager (1996), 5. Int. Conf. on CFB, May 28-31, Beijing, China.

[25] Theologos, K.N., Markatos, N.C. (1992), Trans. IChemE., vol.70, part A, pp.239-245

[26] Theologos, K.N., Markatos, N.C. (1993), AIChE Journal, vol.39, no.6, pp.1007-1017

[27] Gidaspow, D., Therdthianwong, A. (1993), Proc. of 4th Int. conference on circulating fluid beds, pp. 351-358

[28] Samuelsberg, A. (1994), Dr.Ing. Thesis, Telemark Inst. of Technology.

[29] Samuelsberg, A., B.H. Hjertager (1995), Advances in Multiphase Flow - 1995, Elsevier, Amsterdam, pp. 679-688.

[30] Hjertager, B.H., Samuelsberg, A. (1992), KONA Powder and Particle, No. 10, pp. 96-103.

[31] Samuelsberg, A., B.H. Hjertager (1996), 5th Int. Conference on circulating fluidized beds, May 28-31, Beijing, China.

[32] Spalding, D.B. (1980), Recent Advances in Numerical Methods in Fluids, Pineridge Press, pp. 139-168.

[33] Hjertager, B.H. (1998), "Computational fluid dynamics (CFD) analysis of multiphase chemical reactors", in Trends in Chemical Engineering, Vol. 4, pp 45-92 (1998).

[34] Deardorff, J.W. (1971), Journal of Computational Physics 7. pp 120-133

[35] Chapman, S., Cowling, T.G. (1970), "The mathematical theory of non-uniform gases.", 3. edition. Cambridge Univ. press

[36] Lun, C.K.K, Savage, S.B., Jeffrey, D.J., Chepurniy, N. (1984), J of Fluid Mech., vol.140, pp.223-256

[37] Geldart, D., ed, (1986) "Gas Fluidization Technology", J. Wiley, Chichester.

[38] Ouyang, S., Lin, J., Potter, O.E. (1993), Powder Technology, vol.74, pp.73-78

[39] Ouyang, S., Potter, O.E. (1993), Proc. of 4th Int. conference on circulating fluid beds, pp.422-427

[40] Miller, A., Gidaspow, D. (1992), AIChE journal, Vol. 38, no.11, pp.1801-1815.

[41] Azario, E., Tadrist, L., Santinini, R., Pantaloni, J. (1995), Proc. of the 2nd ICMF - '95 - Kyoto, April 3-7, Kyoto, Japan.

[42] Wu, X.L., Maløy, K.J., Hansen, A., Ammi, M, Bideau, D. (1993), Phys.Rev.Lett., Vol 71, pp 1363-1366.

[43] Jenkins, J.T., Savage, S.B. (1983), Journal of Fluid Mechanics, Vol.130, pp. 187-202

Particle Fluctuation Velocity in Gas Fluidized Beds - Fundamental Models Compared to Recent Experimental Data

G.D. Cody
Exxon Corporate Research, Annandale, NJ 08801
S. K. Kapbasov
Karaganda State University, Karaganda, 470070 Kazakhstan
Y. A. Buyevich
University of California, Santa Barbara, CA 93117

We compare the results of two independent measurements of the mean squared particle fluctuation velocity near the wall of gas fluidized beds consisting of monodispersed glass spheres whose diameters, D, span the Geldart A/B transition, at D ≈120μm both sets of experimental data with recent fundamental models for the mean square fluctuation velocity as a function of sphere diameter and superficial gas velocity. These new theoretical and experimental results suggest that the initial fluidized state of monodispersed glass spheres in the Geldart A regime, which has the Geldart Descriptor: "fluidizes before bubbling", may be primarily determined by shear fluctuation forces on these small particles resulting from fluctuations in the circulation flow that establishes itself within the bed and whose magnitude is highly sensitive to both bed geometry and dimensions. This result, if validated by further research, will impact the theoretical and experimental investigation of the gas fluidized state for all particles whose diameter and density place them in the Geldart A regime, from the catalyst particles of petrochemical and refining interest to the monodispersed glass spheres examined in this paper.

INTRODUCTION

The mean squared fluctuation velocity of particles in the dense phase of gas fluidized beds and transfer lines continue to be of great theoretical interest due to the critical role it plays in the dynamics[1] of these complex systems. Given the theoretical interest and the continued industrial importance of fluidized beds in the refining, petrochemical and utility industries, experimental data on the mean squared particle fluctuation velocity, or granular temperature, T*, is surprisingly limited. Indeed, the first direct measurements of the mean squared fluctuation velocity near the wall of gas fluidized beds of monodispersed glass spheres, as a function of sphere diameter and gas flow have only recently been made by two novel, non-intrusive, experimental techniques: Acoustic Shot Noise (ASN) excitation of the wall of the fluidized bed, specifically, the measurement of the average acceleration power spectrum of the wall in a frequency range dominated by random particle impact by Cody et al [2, 3, 4, 5], and Diffusing Wave Spectroscopy (DWS) of the reflection of laser light from the glass spheres, specifically, the analysis of multiple scattering of laser light at one location on the wall to measure the relative motion of the glass spheres near the wall, by Menon and Durian[6].

In this paper we compare the data on the fluctuation velocity obtained in these measurements with each other, as well as with recent first principles models of the granular temperature due to Buyevich[7, 8], and Buyevich and Kapbasov[9].

RECENT EXPERIMENTAL DATA ON PARTICLE FLUCTUATION VELOCITY

As indicated schematically in Fig. 1, the measurements of the fluctuation velocity normal to the wall, v_n, by Cody et al, utilizing Acoustic Shot Noise (ASN) and the measurements of the fluctuation velocity, ∂V, near the wall by Menon and Durian using Diffusing Wave Spectroscopy (DWS) differ in: (a) ***Technique:*** ASN utilizes structure borne sound - DWS utilizes the scattering of coherent light; (b) ***Localization:*** ASN averages over the wall of the fluid bed and measures particle impact at the wall - DWS measures at a point and averages over 7-10 particles near the wall; (c) ***Wall Geometry:*** ASN employed aluminum and glass ***circular*** cylinders of inner diameter ≈ 7.5 cm - DWS employed a glass ***rectangular*** structure of square side 6 cm.

Figure 1
Geometry of Glass Vessel in the Two Experiments

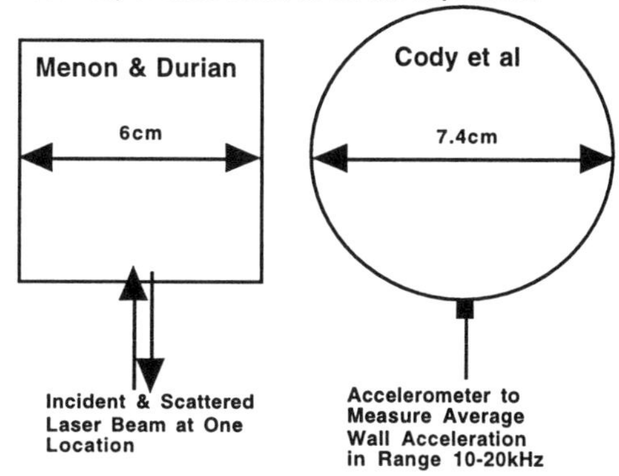

Figure 2
Fluctuation Velocity at Gas Superficial Velocity, $U_s=2U_{mf}$, for Mono-Dispersed Glass Spheres: Cody et al (1996-7), Menon & Durian (1997)

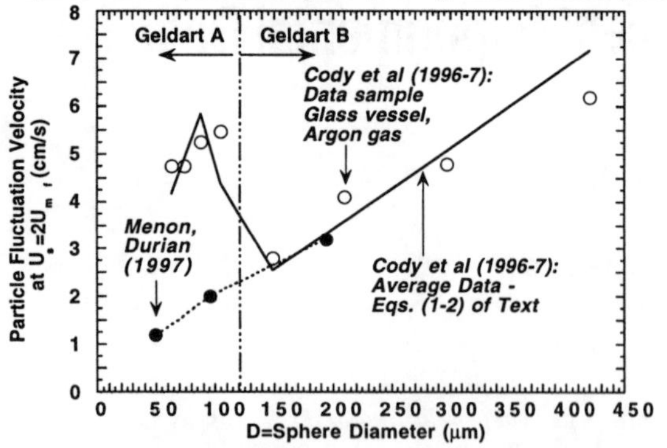

Figure 2 exhibits the fluctuation velocity obtained from the two measurements at a superficial gas velocity, U_s, corresponding to twice the minimum fludization velocity, U_{mf}, or $U=(U_s/U_{mf})=2$, a value chosen to focus on experimental data well into the fluidized state that is independent of the "first-order phase transition" from the fixed bed to the fluidized state. Two sets of data are shown for the ASN measurements of Cody et al. One set, the "Average" fluctuation velocity at the wall, $v_n(D,U)$, at $U\approx2$ is well described for both Geldart A and Geldart B spheres[10, 11] by

$$v_n(D,U=2)=(2U_{mf})\left(\frac{D_o}{D(\mu m)}\right) \qquad (1)$$

where D_o =120µm for D≥149µm (Geldart B), D_o=190µm for D=105µm (Geldart A/B), and D_o=300µm for D≤88µm (Geldart A) [3, 4, 5]. For 1.5<U<8, the scaling with D given by Eq. (1) continues, but for Geldart A and Geldart A/B, $[v_n(D,U)/U_s]$ falls with increasing U indicating the on-set of bubbling; for Geldart B, $[v_n(D,U)/U_s]$ is constant. In general, the ASN data shows that $[v_n(D,U)/U_s] = [G_i(U)/D]$ for all the Geldart regimes with the function G_i (U) different in each. As shown in the above references, the "average" minimum fluidization velocity U_{mf}, is experimentally well fit by

$$U_{mf}(cm/s)=L_i\, D(\mu m)^2 \qquad (2)$$

where for 63µm ≤D≤105µm, $L_i=L_{GA}$ = 1.117×10^{-4} [(cm/s)/(µm^2)], and for 149µm ≤D≤420µm, $L_i=L_{GB}$ = 0.707×10^{-4}[(cm/s)/(µm^2)]. The quadratic variation in U_{mf} with sphere diameter is expected from the Ergun equation[12], and the magnitudes of L_i are consistent with a voidage at minimum fluidization, e_{mf}=0.433 (0.380) for D<(>) 105µm, or a Richardson-Zaki Number[13], n= 4.85 (4.70) for D<(>) 105µm. Thus from Eq. (2) for all the ASN data, $v_n(D,U)$ = $D[UL_iG_i(U)]$ for 1.5<U<8.

The discrete points shown in Fig. 2, for the ASN measurements, are a data sample for the glass cylinder with argon fluidization, for glass spheres with D=63, 74, 88, 105, 149, 210, 297, 420µm and give an indication of experimental deviations from the fit defined by Eqs. (1-2). These points include data from the aluminum cylinder and glass cylinder, with both argon and helium fluidizing gas [2].

For the Geldart B glass spheres[11] (D≥150µm), the agreement between the two ASN and DWS experimental data is remarkable. However for the Geldart A glass spheres (D≤105µm), there is a systematic and interesting difference between the two measurements. The DWS measurements, of Menon and Durian, do not exhibit the bifurcation in the fluctuation velocity observed in the ASN measurements of Cody et al below the Geldart B/A transition at D≈ 120µm. The unexpected bifurcation in the ASN data has been shown to quantitatively meet the one-dimensional first order stability criteria for the uniform fluidized state[3, 4, 5], suggesting that it is a cause, and not a consequence of Geldart A fluidization. It is unlikely that the difference between the experiments is an experimental artifact, given the agreement between the two experiments in the Geldart B regime, and given the fact that both experiments utilize monodispersed glass spheres supplied by the same manufacturer (1600 Series, Spacer Grade Microbeads, 90% true, maximum variation of 6% from average diameter, Cataphote Inc., Jackson. Miss.)

If the difference exhibited between the two experiments in Fig. 2 for the Geldart A regime, is not an experimental artifact, is it only a "wall effect" ? Despite the agreement between the Menon and Durian data and the Cody et al data in the Geldart B regime, it might be argued that the difference between the two experiments in the Geldart A regime might be related to the fact that the optical probe is a measurement at one location at the wall while the ASN measurements average over the wall.

Figure 3
1-D, 1st Order, Initial Stability Criteria, $V_w(D,U=2) > U_*(D,U=2)$:
T*(D,U=2) Data of Menon & Durian (1997) - Unstable D≥49µm;
T*(D,U=2) Data of Cody et al (1996-7) - Stable D≤88µµ.

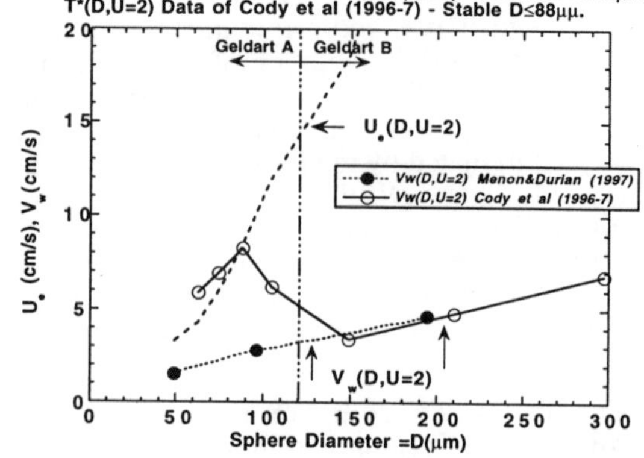

To address this question we consider the initial stability of the Geldart A phase, which has been assigned the Geldart A descriptor of "fluidizes before bubbling". The first order stability theory of Jackson and Wallis[14, 15, 16, 17] for the uniform fluidized state, introduces two critical velocities: the dynamic (acoustic) wave velocity $V_w(D,U)=$ $(T^*(D,U)/(1-e_s)^{0.5}$ and the continuity (voidage) wave velocity $U_e(D,U)=U_t(D)\ ne_s^{n-1}(1-e_s)$ where T*(D, U) is the granular temperature or square of the fluctuation velocity, v_n, , and $U_t(D)$ is the Stokes velocity[3, 4, 5]. In the Jackson theory if $V_w(D,U)>U_e(D,U)$, the uniform fluidized bed is stable against 1-D fluctuations in voidage and should not exhibit bubbling just above minimum fluidization gas velocity. In Fig. 3 we plot the two critical velocities at U=2, for the monodispersed glass spheres used in the experiments of Cody et al and Menon and Durian. From Fig. 3 we note that the uniform state of the "Geldart A spheres" of Menon and Durian, in contrast to the "Geldart A spheres" supplied by the same manufacturer to Cody et al, are unstable with respect to small perturbations in voidage in Jackson's model. We next show that there is independent experimental evidence for such an instablity in the Menon and Durian data by comparing bed expansion obtained in the two experiments.

Figure 4

Initial bed expansion, α(D,U) for monodispersed glass spheres: Menon and Durian (1997) and Cody et al (1996-97)

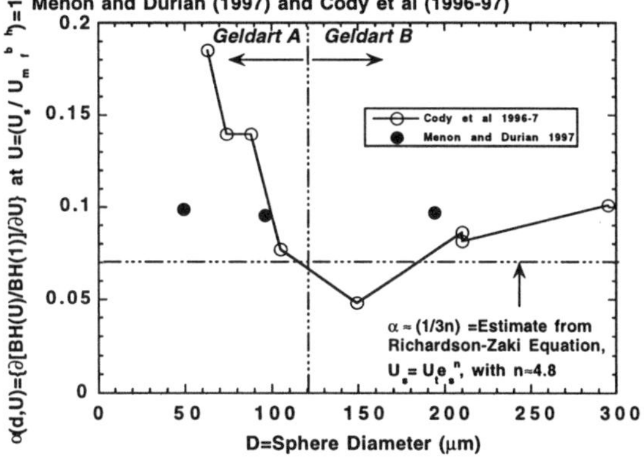

In Figure 4, we compare, for the two experiments, the slope of the initial bed expansion which is defined as $\alpha(D,U)=[\partial \ln BH(D,U)/\partial U]$ at U=1, where BH(D,U) is the bed height for spheres of diameter D, at gas flow $Us=(U)(U_{mf})$. We note that α(D,U) is a "volume" measurement of T* compared to the "near wall" ASN and DWS measurements of T*. We note in Fig. 4 that α(D,U=1) for the "Geldart A spheres" of Menon and Durian, has the same magnitude as α(D,U=1) for their "Geldart B spheres". In contrast, we note in Fig. 4 for the experiments of Cody et al, that the α(D,U=1) for the "Geldart A spheres" is about of 2 times larger than the magnitude of α(D,U=1) for their "Geldart B glass spheres". Both results are consistant with the stability predictions exhibited in Fig. 3.

Thus from Figs. 2-4, it would appear that the "Geldart A glass spheres" in the experiments of Menon and Durian exhibit throughout the bed, not just at the wall, "Geldart B fluidization" - a conclusion derived from: (1) the magnitude of the fluctuation velocity, (2) the 2-wave stability criteria and (3) the slope of the initial bed expansion. In the next section we examine recent fundamental models for the mean squared fluctuation velocities of monodispersed glass spheres which also support this surprising conclusion. Indeed these models suggest that it is the strong difference in bed geometry for the two experiments exhibited in Fig. 1, and the consequent difference in circulation flow and shear fluctuation forces within the vessels, that is responsible for this surprising difference in the magnitude of the particle fluctuation velocity obtained in the two experiments for monodispersed glass spheres with D≤100μm.

RECENT FUNDAMENTAL THEORIES OF THE PARTICLE FLUCTUATION VELOCITY

The first model for the data of Cody et al exhibited in Fig. 2, was due to Buyevich[7] who developed a first principles statistical model for the mean squared fluctuation velocity, based on random impacts due to "pseudoturbulent fluctuations" induced in particle velocity, no energy loss per collision, critical but straightforward assumptions about the form of the random impact force, the condition that the fluctuation velocity is isotropic so that the granular temperature is well defined by $T^*(D,U) \equiv v_n(D,U)^2$, and finally, the neglect of shear fluctuation forces. A major unknown in these "pseudoturbulent" models[18] is the concentration fluctuation variance $\langle\phi'^2\rangle$, and the early models of Buyevich used the Carnahan and Starling equation[19]

$$\langle\phi'^2\rangle \equiv \langle(\phi-\langle\phi\rangle)^2\rangle = \phi^2\left(1+2\phi\left(\frac{4-\phi}{(1-\phi)^4}\right)\right)^{-1} \quad (4)$$

where in the last expression for $\langle\phi'^2\rangle$, $\phi \equiv \langle\phi\rangle$.

Figure 5

Experimental Particle Fluctuation Velocity and Buyevich (1997) Model: No Free Parameters, Impact Fluctuations, No Collisional Energy Loss, Concentration Fluctuation Variance, <φ'²> by Eq. (4).

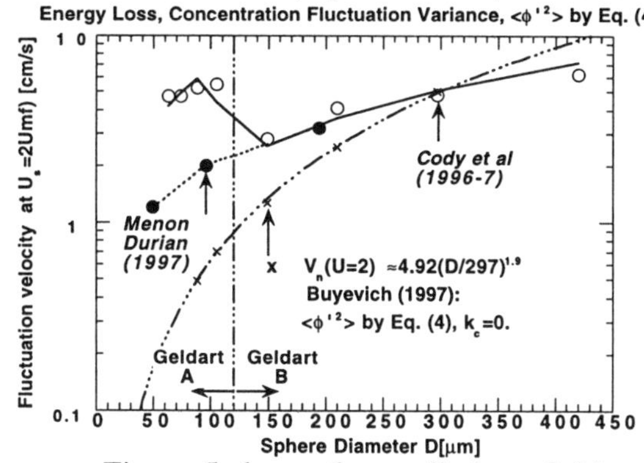

Figure 5 shows the prediction of this model for the fluctuation velocity, ***with no free parameters***. The model is in rough agreement with the Geldart B data, but falls considerably below the experimental data for Geldart A glass spheres. The model exhibits D^2 dependence on sphere diameter for U constant, rather than the experimental D dependence exhibited in Fig. 2, by the Geldart B glass spheres of Cody et al [see Eqs.(1,2)] and both the Geldart A and B glass spheres of Menon and Durian.

Recently, Buyevich [8] introduced a model to account for the bifurcation across the Geldart A/B transition shown for the ASN data of Fig. 2. The model includes random shear-induced fluctuations of particles in addition to the pseudoturbulent fluctuations of reference [7] . The theory is one dimensional and has two free parameters. One is the shear flow which sets the magnitude of the peak shown in Figure 6, the other a length scale, S_o, which sets the width and location of the peak. As shown in Figure 6, a simple one-dimensional version of model captures the rapid change in the fluctuation velocity for D≤105μm found by Cody et al through the introduction of a circulation flow that produces a random shear stress through its interaction with random particle motion. As noted in the paper: ***"The mean squared fluctuation velocity in a fluidized bed of relatively small particles in this model is primarily determined by properties of a circulation that establishes itself with the bed. These properties would be highly sensitive to the bed geometry and dimensions".*** The absence of a bifurcation in the DWS measurements of Menon and Durian, would then be ascribed to the striking difference in the geometry illustrated in Fig. 1, and the subsequent difference in circulation flow for the ASN and DWS experiments. The conjecture that Geldart A fluidization is dependent on circulation within the fluidized bed, if correct, has significant implications for all experimental and theoretical research on all particles whose diameter and density place them in the Geldart A fluidization regime!

Figure 6

Fluctuation Velocity at $2U_{mf}$ for Monodispersed Glass Spheres: Cody et al (1996-7), Menon & Durian (1997) Compared to Model of Buyevich &Cody (Buyevich & Cody(1998))

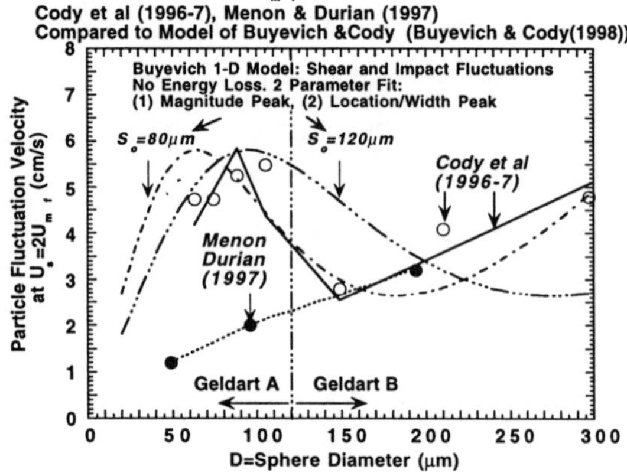

Buyevich and Kapbasov[9] recently extended the Buyevich model exhibited in Fig. 5. Again it is a first principles statistical model for the mean squared fluctuation velocity based on: (1), critical but straight forward assumptions about the form of the random impact force; (2), the condition that the fluctuation velocity is isotropic; (3), the absence of shear fluctuation forces; (4), the inclusion of collisional energy loss as a free parameter. As will be seen in Figure 7, the remarkable agreement of the Buyevich Kapbasov model predictions ***with all of the DWS data and the Geldart B ASN data*** supports the hypothesis that the "Geldart A glass spheres" of Menon and Durian are behaving like "Geldart B glass spheres" due to low shear circulation at the face of the square vessel, and the subsequent reduction in the magnitude of the shear fluctuation forces.

In the new Buyevich Kapbasov model, the concentration fluctuation variance $<\phi'^2>$, is given by a formula first introduced by Buyevich[20] in the early seventies for highly concentrated systems where $\phi \equiv <\phi>$ has the maximal value, $\phi_* = 0.60\text{-}0.65$, attributed to a hypothetical state of close packing

$$<\phi'^2> = \phi^2\left(1 - \left(\frac{\phi}{\phi_*}\right)\right) \qquad (5)$$

Eq. (4), which is used in the Buyevich model displayed in Fig. 5, follows from the thermodynamic theory of fluctuations and the approximate hard sphere statistical physical model of Carnahan and Starling[19]. In contrast, Eq.(5) is a generalization of Smoluchowski's colloidal statistics to dense concentrated colloids[20]. It can not be known, a priori, which model is appropriate for fluidized beds, and the choice between Eqs. (4) and (5) can only be based on experiment. From the experimental data summarized in this paper for gas fluidized beds, and recent experimental data on particle pressure in solid-liquid systems by Zernit et al[21], Buyevich[22] has shown that Eq. (4) underestimates the concentrational fluctuation variance, and hence the granular temperature, and that Eq. (5) should be preferred for modeling, by the pseudoturbulent approach[18], both gas and liquid fluidized beds.

Figure 7

Experimental Particle Fluctuation Velocity and Buyevich and Kapbasov (1998) Model: Impact Fluctuations, Collisional Energy Loss, k_c, Concentration Fluctuation Variance, $<\phi'^2>$ by Eq. (5).

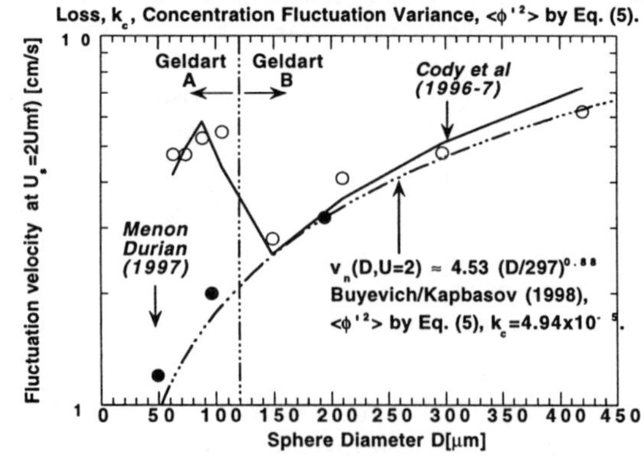

The recent Buyevich Kapbasov model[9] employing Eq. (5) neglects random shear fluctuations, so it is only expected to apply in the Geldart B regime. As noted in the Figure 8, it also introduces a free parameter, k_c , the fractional energy loss per collision. As indicated in Figure 7 the remarkable agreement between the new model and the Geldart B spheres of Cody et al and both the Geldart B and Geldart A spheres of Menon and Durian occurs for $k_c = 4.94 \times 10^{-5}$. This value for k_c corresponds to a coefficient of restitution, $e_p = 0.9999$, which is physically realistic since the fluctuation velocities for the glass spheres are well below the velocities required to introduce inelastic collisions[23] in hard materials (≈0.1m/s) due to plastic deformation of Hertzian contacts[24, 25, 26]. The small difference from $e_p = 1$

only includes the effect of energy losses due to surface friction, and the work of viscous stresses that arise as a result of jump-like change of the fluid velocity in the vicinity of colliding particles.

Figure 8

Experimental Particle Fluctuation Velocity and Buyevich and Kapbasov (1998) Model: Impact Fluctuations, Concentration Fluctuation Variance, $<\phi'^2>$ by Eq. (5), Energy Loss, k_c=0, 4.94x10⁻

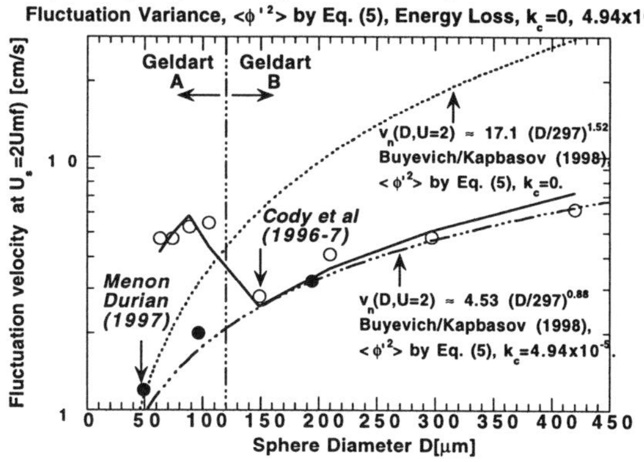

Figure 8 exhibits the effect of a finite value for k_c in terms of the new Buyevich and Kapbasov model, and we note that the introduction of the small but finite collisional energy loss leads to a drop in the magnitude of $v_n(D=300\mu m,U=2)$ by a factor of about 7, corresponding to a reduction in the granular temperature, $T^*(D=300\mu m,U=2)$ by a factor 50. The high sensitivity of the granular temperature to such a small change in energy loss per collision can only be ascribed to the increased correlation in particle velocity after multiple collisions due to a finite, but not zero value for k_c, which reduces the mean squared fluctuation velocity, $T^*(D,U_s)$. A similar effect has been in simulations of dense gas models for granular systems in a recent thesis of Bizon[27].

Of equal significance in Fig. 8 is the signficant change in the power law dependence of the fluctuation velocity at U=2, from $v_n(D,U=2) \approx D^{1.5}$, with no energy losses per collision, to $v_n(D,U=2) \approx D^{0.9}$, with finite, but essentially zero collisional energy loss. As noted earlier in the discussion following Eqs. (1,2), the scaling $[v_n(D,U)/U_s] = [G_i(U)/D]$ is found for all the Geldart regimes in the ASN data, where the function $G_i(U)$ is different for each Geldart regime[3, 4, 5]. As also shown in these publications, the linear scaling of $v_n(D,U=2)$ with D in the Buyevich and Kapbasov model arising from the inclusion of collisional energy loss can be readily understood dimensionally by a Langevin model for the random particle vector velocity. We first note that from Eq.(1), for $U_s=2U_{mf}$, $v_n(D,U_s) = U_s(D_o/D)$ and hence $T^*(D,U_s) = U_s^2(D_o/D)^2$. If τ_p is the Langevin relaxation time, it is easily shown[2, 3, 4, 5] that the power $q_c(D,U_s)$ per unit volume required to keep $T^*(D,U_s)$ constant is given by:

$$q_c(D,U_s) = \left(\frac{3\rho_m T^*(D,U_s)}{\tau_p}\right) \tag{6}$$

where the quantity, ρ_m, is defined by $\rho_m = \rho_o\,(1\text{-}e_s)$.

The power dissipation given in Eq. (6) is supplied by the fluidizing gas. It is also readily shown[2] from the Ergun equation[12] that the power per unit volume, to drive gas of viscosity μ_g through a "fixed bed" of voidage, e_s, at a gas superficial velocity, U_s, is

$$Q(D,U_s) = (150)\left(\frac{(1\text{-}e_s)^2}{e_s^3}\right)\mu_g\left(\frac{U_s^2}{D^2}\right) \tag{7}$$

If we assume that $q_c(D,U) \approx K\,Q(D,U_s)$ where K is a dimensionless constant which is a function of the voidage in the bed. Then from Eqs. (6,7):

$$T^*(D,U_s) = (50K)\left(\frac{(1\text{-}e_s)}{e_s^3}\right)\left(\frac{\mu_g\tau_p}{\rho_o}\right)\left(\frac{U_s^2}{D^2}\right) \tag{8}$$

which defines the dimensional dependence of scaling parameter D_o, of Eq. (1) on the mass density of the particle, ρ_o, the viscosity of the fluidizing gas, μ_g, and the Langevin relaxation time, t_p

$$D_o^2 \equiv (50K)\left(\frac{(1\text{-}e_s)}{e_s^3}\right)\left(\frac{\mu_g\tau_p}{\rho_o}\right) \tag{9}$$

The quantity k_c is defined in the Buyevich and Kapbasov[9] paper by

$$q_c(D,Us) = k_c\left(\frac{3\rho_m T^*(D,Us)}{T_l}\right) \tag{10}$$

where T_l is the collision time, given in the dense gas kinetic theory by

$$T_l(D,Us) = \left(\frac{D}{v_n}\right)\left(\frac{\pi^{0.5}}{24\phi\chi(\phi)}\right) \tag{11}$$

and $\chi(\phi)=[(1\text{-}0.5\phi)/(1\text{-}\phi)^3]$ is the Enskog factor that describes the decrease in collision time in a dense gas of identical spheres as compared to a dilute gas of the same spheres[7, 8]. From Eqs. (6, 10), $k_c = T_l/\tau_p$ and from the Langevin equation applied to one collision[2], $k_c = [1\text{-}e_p^2]/4$. From Eq. (1) and Eq. (11) and the Richardson Zaki number of 4.70 found experimentally, we find $T_l(D,U=2) \approx 9.5\times10^{-5}$ s. From the value, $k_c=4.9\times10^{-5}$, used to fit the data in Figs. 7 and 8, we obtain $\tau_p \approx 1.9$s, about a factor of 3-6 times slower than the measured bed collapse times for Geldart A and B monodispersed glass spheres[2].

SUMMARY

Recent theoretical models for the granular temperature of glass spheres dominated by random impact and small energy losses per collision with and without shear fluctuation forces due to Buyevich and co-workers[8, 9, 22] when compared with recent independent measurement of the granular temperature[2, 4, 6] support the hypothesis that the Geldart A fluidization regime for monodispersed glass spheres with sphere diameter $D \leq 100\mu m$ depends of the strength of random shear fluctuations of the gas circulation and thus depends strongly on bed geometry and shear circulation. This result, if confirmed by further experiment and theory, impacts fundamental theories of the stability of the uniform fluidized state as well as the interpretation of recent measurements of such fundamental properties of the fluidized state such as the granular pressure[28, 29], and the granular viscosity and sound velocity[30].

NOTATION

ASN = Acoustic Shot Noise measurement of fluctuation velocity v_n
BH(D,U)= Bed Height as a function of D and $U=U_s/U_{mf}$
D_o = Geldart regime dependent scaling for sphere diameter in Eq. (1) in microns [μm]
D = D(μm) = Particle diameter in microns [μm]
e_{mf} = Voidage of dense phase at U_{mf}
e_s = Voidage of dense phase at U_s
e_p = Coefficient of restitution for glass sphere
DWS = Diffusing Wave Spectroscopy measurement of fluctuation velocity ∂V
$G_i(U)$ = Geldart regime dependent function of U that defines $[v_n(D,U)/U_s] = [G_i(U)/D]$
k_c = Fractional energy loss per collision
L_i = Geldart regime dependent constant defined by $U_{mf} = L_i\, D(\mu m)^2$ with units $[(cm/s)/(\mu m)^2]$
n = Richardson-Zaki number defined by $U_s=U_t e_s{}^n$
$q_c(D,U_s)$ = Power per unit volume required to maintain $T^*(D,U_s)$ constant in fluidized bed
$Q(D,U_s)$ = Power per unit volume to drive gas of viscosity μ_g through fixed bed of voidage, e_s
$T^*(D,U)$ = Granular Temperature $\equiv v_n{}^2(D,U)$
T_I = Collision time for glass spheres [s]
$U = (U_s/U_{mf})$
U_s = Gas superficial velocity [cm/s]
U_{mf} =Gas superficial velocity at fluidization on-set [cm/s]
$U_e(D,U)$ = "Continuity Velocity" of 2-Wave Stability Theory [cm/s]
U_t = Stokes velocity $=(\rho_o g D^2/18\mu_g)$ [cm/s]
$v_n(D,U)$ = Fluctuation velocity from ASN data: normal to the wall and averaged over wall [cm/s]
$V_w(D,U)$ = "Dynamical Velocity" of 2-Wave Stability theory [cm/s]

Greek Symbols

$\alpha(D,U=1)$= $[\partial \ln BH(D,U)/\partial U]$ at U=1, normalized iniital slope of bed expansion
∂V = Fluctation velocity in DWS data, an average over ≈ 9 sphere diameters of wall [cm/s]
ΔT_c = Bed collapse time [s]
ρ_o = Mass density of glass sphere[gm/cm^3]
ρ_m = Mass density of dense phase $=\rho_o(1-e_s)$
ϕ = Particle concentration in dense phase=$(1-e_s) \equiv \langle\phi\rangle$
ϕ' = Fluctuation in particle concentration in dense phase
$\langle\phi'^2\rangle$ = Concentration fluctuation variance
τ_p = Langevin relaxation time [s]
μ_g = Viscosity of fluidizing gas [poise]

References

1. R. Jackson, in *Fluid Particle Technology, AIChE Symposium Series 301*, A. W. Weimer, Ed., vol. 90, American Institute of Chemical Engineers, New York, 1994, pp. 1-30.
2. G. D. Cody, D. J. Goldfarb, G. V. Storch Jr., A. N. Norris, *Powder Technology*, **87**, (1996) 211-232.
3. G. D. Cody, D. J. Goldfarb, in *AIChE Symp. Ser.: Progress in Fluidization and Fluid Particle Systems*, D. King, Ed., vol. 93, AIChE, New York, 1997, pp. 18-24.
4. G. D. Cody, D. J. Goldfarb, in *Fluid.-IX*, L.-S. Fan, T. M. Knowlton, Eds., Eng. Found., New York, 1998, pp. 53-60.
5. G. D. Cody, D. J. Goldfarb, in *Dynamics in Small Confining Systems-III*, M. Drake, J. Klafter, R. Kopelman, Eds., vol. 464, Materials Research Society, Pittsburgh, Pa, 1997, pp. 325-338.
6. N. Menon, D. J. Durian, *Phys. Rev. Lett.*, **79**, (1997) 3407-3410.
7. Y. A. Buyevich, *Chem. Engng. Sci.*, **52**, (1997) 123-140.
8. Y. A. Buyevich, G. D. Cody, Particle fluctuations in homogeneous fluidized beds, Paper 207-World Congress on Particle Technology-3, Brighton, UK (1998).
9. Y. A. Buyevich, S. K. Kapbasov, *Int. J. Fluid Mech. Res.*, **26**, (1999) 72-97.
10. D. Geldart, *Powder Technology*, **7**, (1973) 285-282.
11. D. Geldart, in *Gas Fluidization Technology*, D. Geldart, Ed., John Wiley, New York, 1986, pp. 11-51.
12. D. Kunii, O. Levenspiel, *Fluidization Engineering*, Krieger, Malabar, Florida, USA, 1987, p. 73.
13. J. P. Couderc, in *Fluidization*, J. F. Davidson, R. Clift, D. Harrison, Eds., Academic Press, New York, 1985, pp. 1-44.
14. R. Jackson, *Trans. Instn. Chem. Engrs.*, **41**, (1963) 13-21.
15. G. B. Wallis, *One-Dimensional Two-Phase Flow*, McGraw-Hill, New York, 1969, p. 122-242.
16. R. Jackson, in *Fluidization*, J. F. Davidson, R. Clift, D. Harrison, Eds., Academic Press, New York, 1985, pp. 47-72.
17. M. Nicolas, J.-M. Chomaz, E. Guazzelli, *Phys. Fluids*, **6**, (1994) 3936-3944.
18. Y. A. Buyevich, S. K. Kapbasov, in *Multiphase Reactor and Polymerization System Hydrodynamics*, N. Cheremisinoff, Ed., Gulf, Houston, 1996, pp. 119-166.
19. N. F. Carnahan, K. E. Starling, *J. Chem. Phys.*, **51**, (1969) 635-637.
20. Y. A. Buyevich, *Chem. Engng. Sci.*, **26**, (1971) 1195-1201.
21. R. Zernit, M. L. Hunt, C. E. Brennen, *J. Fluid Mech.*, **353**, (1997) 261-283.
22. Y. A. Buyevich, *Ind. Eng. Chem. Res.*, **38**, (1999) 731-743.
23. D. Gidaspow, *Multiphase Flow and Fluidization - Continuum and Kinetic Theory Descriptions*, Academic Press, San Diego, 1994, p. 239-354.
24. K. L. Johnson, *Contact Mechanics*, Cambridge University Press, Cambridge, 1986, p. 353.
25. S. F. Foerster, M. Y. Louge, H. Chang, K. Allia, *Phys. Fluids*, **6**, (1994) 1108-1115.
26. N. V. Brilliantov, F. Spahn, J. Hertzsch, T. Poschel, *Phys. Rev. E*, **53**, (1996) 5382-5392.
27. C. A. Bizon, *Simulations of Wave Patterns in Oscillated Granular Media*, , Thesis, Department of Physics, University of Texas at Austin (1998).
28. C. S. Campbell, K. Rahman, *Meas. Sci. Tech.*, **3**, (1992) 709-712.
29. W. Polashenski Jr., J. C. Chen, *Powder Technology*, **90**, (1997) 13-23.
30. D. Gidaspow, H. Lu, in *Fluid.-IX*, L.-S. Fan, T. M. Knowlton, Eds., Eng. Found., New York, 1998, pp. 661-668.

On the Shape of Grade Efficiency Curves Computed for Centrifugal Dedusters by CFD with Lagrangian Particle Tracking

Alex C. Hoffmann and Attilio Berrino
Department of Chemical Engineering, University of Groningen, Nijenborgh 4, 9747 AG Groningen, The Netherlands

Madhumita Bhowmick Ray
Department of Chemical Engineering, National University of Singapore, 10 Kent Ridge Crescent, Singapore 119260

The factors influencing the shapes of cyclone grade efficiency curves generated by Computational Fluid Dynamics (CFD) and Lagrangian particle tracking are studied. Two software packages SPRINT and FLUENT were used for a complete picture. Computed curves both with and without the effect of turbulent particle dispersion are compared with experimental ones. It is found that the steepness of the predicted curves without turbulent dispersion agrees well with experimental curves. Including turbulent dispersion in the particle tracking results in the formation of a 'tail', but the shape and the position do not agree well with the 'duck-tails' in the experimental curves. These features are discussed.

Cyclone separators are probably the most widely used dedusters in broad range of industries including powder processing in chemical, food and drug industries. A large amount of literature describes the complexity of the fluid and particle dynamics in the otherwise structurally simple cyclones. The flowpattern of the fluid in the cyclone body was studied experimentally (1-4), analytically (5, 6) and numerically (7, 8). The majority of the works on flow pattern in cyclone concentrated mainly on reverse flow cyclones with tangential gas inlets and their application to other cyclone configurations remains uncertain. Recent development of high performance computers and numerous CFD packages paved the way for simulation of gas flow pattern in the cyclone and provided a better approach in understanding the gas flow dynamics in different configurations of cyclones. Lately, CFD was successfully adopted by several researchers for different cyclone uses including a study on predicting the performance of cyclone samplers (9), a cyclone used for rotary kiln preheater (10), and a cyclone with different design modifications (11).

The separation performance of a fluid-particle separator is characterized mainly by grade efficiency data since the overall efficiency is inadequate as a general measure of the separator. The slope of the grade efficiency curve strongly depends on the particle motion in the cyclone, that can be correlated to cyclone operational and design parameters. Liden and Gudmundsson (12) analysed published collection efficiency data for four different classes of cyclone. The study indicated that parameters such as annular Reynolds number (a Reynolds number based on the inlet velocity and the annular distance between the vortex tube and the cyclone wall) and ratio of cut-off size to cyclone body diameter can explain all variations for all four classes of cyclone at annular Reynolds number range of $0.5\text{x}10^3 - 100\text{x } 10^3$. Additional parameters such as the ratio of vortex tube length to the cyclone body diameter, inlet area and the ratio of cyclone body diameter to the vortex tube inner diameter also influence the cut-off size and the shape of the grade efficiency curve. In the light of limited success of the existing research, the authors indicated the need of additional experiments in order to elucidate the importance of different design parameters on the shape of grade efficiency curves. Numerical simulation provides the immense opportunity to study the effects of these design parameters on different classes of cyclone. However, at the onset it is necessary to identify how the computed grade efficiency curves respond to different constraints imposed on the calculation of particle trajectory in the fluid field. Earlier Hoffmann et al. (13) showed that inclusion of a dust collection hopper could improve agreement with experiment in the absolute position of computed grade efficiency curves. The present paper explores the shape of the grade efficiency curves by changing different solution parameters available in the CFD programs.

THEORY AND COMPUTATIONAL METHODS

Calculation of the flow field

Two different CFD programs FLUENT and SPRINT were used to calculate the flow pattern and grade efficiency in a Stairmand high efficiency cyclone with a body diameter of 0.2 m. In both the programs the flow field is calculated in an axisymmetrical pseudo 2-dimensional geometry. SPRINT uses a turbulent modeling scheme hybrid between an algebraic stress model and a full Reynolds stress model (RSM) which has been shown to give realistic flow fields (11, 13) whereas the turbulence modeling in FLUENT was achieved by RSM model. SPRINT uses skew upwind differencing scheme (SUDS) and FLUENT offers a power-law, second order upwind or quadratic upwind differencing (QUICK) schemes to use for interpolation between grid points and to calculate the derivatives of the flow variables. The set of discretized algebraic equations is solved by semi-implicit iterative scheme.

Particle tracking

The particle tracking in both the programs were performed in a Lagrangian frame of reference. The prerequisite of using a Lagrangian frame in a pre-calculated flow field is that the dispersed phase is sufficiently dilute and the particle streams are injected into the continuous phase with a well defined entrance and exit conditions. Particle tracking is done by solving the particle equations of motion in cylindrical coordinates (r, θ, z):

$$\frac{du_z}{dt} = -\alpha(u_z - v_z - v_z');$$

$$\frac{du_r}{dt} = \frac{u_\theta^2}{r} - \alpha(u_r - v_r - v_r');$$

$$\frac{du_\theta}{dt} = -\frac{u_r u_\theta}{r} - \alpha(u_\theta - v_\theta - v_\theta') \quad (1)$$

$$\alpha = \left(\frac{18\mu_F}{\rho_P d_P^2}\right)\left(\frac{C_D Re}{24}\right) \quad (2)$$

The drag coefficient, C_D is evaluated from a generic expression:

$$C_D = a_1 + \frac{a_2}{Re} + \frac{a_3}{Re^2} \quad (3)$$

where the constants a_i depend on the particle Reynolds number.

$$Re = \frac{\rho_F d_P |u_z - v_z|}{\mu_F} \quad (4)$$

The equations of motion are integrated over a time interval Δt (calculated from the particle velocity and the size of a computational cell) so small that α can be assumed constant to give a closed solution:

$$u_z = v_z + v_z' + (u_{z,0} - v_z - v_z')\exp(-\alpha\Delta t)$$

$$u_r = v_r + v_r' + \frac{u_{\theta,0}^2}{\alpha r_0} + (u_{r,0} - v_r - v_r' - \frac{u_{\theta,0}^2}{\alpha r_0})\exp(-\alpha\Delta t)$$

$$u_\theta = v_\theta + v_\theta' - \frac{u_{r,0}u_{\theta,0}}{\alpha r_0} + (u_{\theta,0} - v_\theta - v_\theta' + \frac{u_{r,0}u_{\theta,0}}{\alpha r_0})\exp(-\alpha\Delta t) \quad (5)$$

Turbulent particle dispersion is mimicked by adding a stochastic component to the mean gas velocity, consistent with the calculated turbulence intensity. For instance for the axial gas velocity, fluctuation v_z' is:

$$v_z' = \xi\sqrt{<v_z' v_z'>} \quad (6)$$

where $<v_z' v_z'>$ is the Reynolds stress component as calculated during the CFD simulation and ξ is a random variable, normal with mean zero and variance 1. The particle path is then tracked in a gas flow of $v+v_z'$ for the lifetime of an eddy, $\tau = 0.16\ k/\varepsilon$, whereafter a new fluctuating gas velocity is computed. In case of FLUENT, the lifetime for an eddy in RSM model is given as $\tau = 0.30\ k/\varepsilon$ (15).

EXPERIMENTAL

The details of the experimental rig and analytical systems are described elsewhere (16).

Experimental dust was well dispersed in compressed air by passing the dust through a sand-blasting nozzle. In order to ensure full dispersion, a dispersant (Aerosil 200, 1-2% by weight of the dust) was premixed with the experimental dust. The particle size distribution from the different points of the rig was analysed by Joyce-Loebl disc centrifuge.
The cyclone was a Stairmand high efficiency cyclone of 0.2 m body diameter with three different vortex finder diameters, 0.053, 0.075 and 0.1 m. Three different velocities of 10, 15 and 20 m/s were used.

RESULTS AND DISCUSSION

Initially several simulations were performed in order to obtain grid independent solutions for some case files, and the effects of different numerical and solution parameters were also tested. For example, doubling the number of cells from the initial 9880 cells did not change the converged solution in FLUENT. In both simulation packages the inlet turbulence intensity had to be prescribed. Figure 1 shows the effect of inlet turbulent intensity on the grade efficiency of cyclone predicted by SPRINT. The simulations using both packages indicated no change in gas flowpattern by changing inlet turbulent intensity over a wide range, which is also reflected in the grade efficiency as shown in Figure 1.

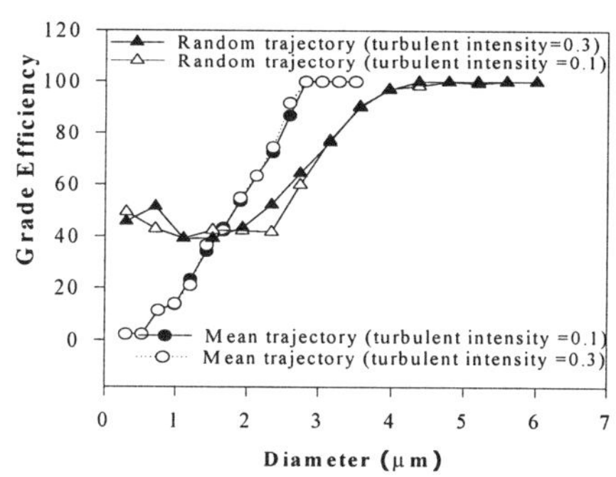

Figure 1, Effect of turbulent intensity on grade efficiency of the cyclone.

Upon identifying the optimum solution techniques for both the programs, they were used to simulate the flow profile in the test cyclone under different operating and design parameters. In Figures 2 and 3, the grade efficiency curves generated by both SPRINT and FLUENT with and without turbulent particle dispersion are compared with experimental data at low solid loadings. A number of factors may influence the shape of the grade-efficiency curves in the physical system, such as: i) the particles are injected at a range of radial positions over the inlet, ii) the inwards radial velocity between outer and inner vortices is axially non-uniform, iii) turbulent dispersion, iv) particle agglomeration, and v) solid loading.

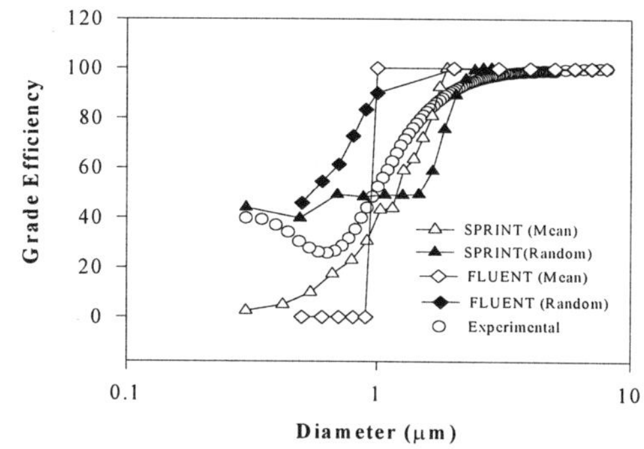

Figure2, Experimental and predicted grade efficiency of cyclone.
(Diameter of the vortex finder=0.075 m, velocity=20 m/s)

Considering first the mean track simulations in which only the mean gas velocity is considered for the particle tracking. These simulations account for factors i) and ii) above at least in the case of SPRINT. In case of FLUENT, the particles were injected from only one radial position in the inlet giving a sharp cut: the simulations only indicate a cut size. It is clear that the steepness or the slope of the experimental curves is well matched by the mean track simulations of SPRINT, perhaps indicating that factors i) and ii) are most important in determining this at low solid loadings.

Turning now to the effect of turbulent dispersion. A feature of the experimental curves is the presence of so-called 'duck-tails' (higher efficiency for smaller particles in the fine range). Such tails are often, but not always, seen, and they do not always begin at the same diameter. For instance, in the study of Bernard et al. (14) the 'duck-tail' began at 2 μm rather than the 0.8 μm in this study. A possible explanation for this feature is turbulent particle dispersion, which is stronger for smaller particles, bringing small particles, that would otherwise be lost, to the cyclone wall to be collected.

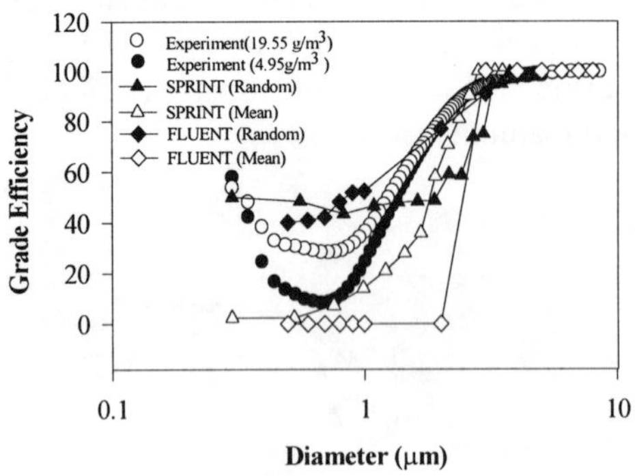

Figure 3, Experimental and predicted grade efficiency of cyclone.
(Diameter of the vortex finder=100 mm, velocity=10 m/s)

The effect of turbulent dispersion was accounted for in both the programs and both exhibit a tail-like feature, though neither the shape nor the position match experiment well. The turbulent dispersion also pushes the rest of the curve to the right, in the case of SPRINT away from the experimental data, but the absolute position may also be influenced by other factors (13). Some of the experimental curves exhibit separation efficiency minima (at around 0.8 μm) much lower that what the random track simulations indicate should be possible (see Figure 3). It is reasonable to ask why the effect of turbulent dispersion in collecting such particles is much less than the simulations indicate. One possibility is that the radial turbulence intensity is overestimated in the CFD flow calculations simulations. Another, more likely, one is that in the physical system turbulent dispersion cannot bring a particle all the way to the wall since transverse turbulence is attenuated near the wall. This is not accounted for sufficiently in the simulations at present.

Without going into detail we note that the shape of the tailing in the experimental curves is more consistent with the notion of small particles being collected as agglomerates than with the effect of turbulent dispersion. Whether these are formed in the cyclone or never been broken up in the feed, in spite of the very efficient dispersion system used is a matter for further study.

The difference in the experimental curves in Figure 3 is due to different solid loadings. The solid loading has a strong effect on the shape of grade-efficiency curves, for which the simulations performed here cannot account. This, and other features will be discussed in a longer article under preparation. Finally it is possible that Brownian particle motion, although the mean motion due to this is on a scale orders of magnitude less than the dimensions of the cyclone, may play a role in bringing small particles through boundary layers to the cyclone walls or to other particles to form agglomerates. Such an effect cannot be accounted for by the sort of simulations described here.

CONCLUSIONS

The shape of grade efficiency curves predicted by two CFD programs SPRINT and FLUENT were compared with experimental curves. For very small particles, the drag calculation included in FLUENT may require correction (15). The general features of the experimental curves were well matched by the mean track simulations when the particles are injected at a range of radial positions over the inlet. Including turbulent dispersion appeared to give rise to a tail-like feature in both the simulation packages, but the shape and position neither agreed between the packages or with the experimental results.

The shape and the slope of the grade efficiency curves determine the effectiveness of a fluid-particle separator, and thus commands greater attention than that has been bestowed in earlier research. Numerical simulation using CFD is emerging as an efficient tool in conducting such studies for wide variety of designs of the centrifugal dedusters like cyclones.

ACKNOWLEDGEMENTS

The authors wish to express their appreciation to Ms. Zhao Wennan for her contribution to the paper by performing several simulations.

NOTATION

u_r, u_z, u_θ particle radial, axial and angular velocity, m/s and rad/s
v_r, v_z, v_θ gas radial, axial and angular velocity, m/s and rad/s
u_r', u_z', u_θ' fluctuations in particle velocity in r, z and θ directions, m/s and rad/s
v_r', v_z', v_θ' fluctuations in gas velocity in r, z and θ directions, m/s and rad/s
C_D drag coefficient
d_p diameter of the particle, m
k kinetic energy of turbulence, $m^2 s^{-2}$
ε rate of dissipation of kinetic energy per unit mass , m^2s^{-3}
μ_F gas viscosity, kg/m.s
ρ_F gas density, kg/m^3
ρ_p density of the particles, kg/m^3

LITERATURE CITED

1. **ter Linden,** "Investigations into Cyclone Dust Collectors", *Proc. I. Mech. E.,* **160**, pp 233-254 (1949).
2. **Shepherd, C.B., and C.E. Lapple,** "Flow Pattern and Pressure Drop in Cyclone Dust Collectors", *Industrial Engineering Chemistry*, **32**, pp1246-1248 (1940).
3. **Mothes H., and F. Loffler**, "The Influence of the Dust Loading on the Gas Flow in a Gas Cyclone" *Proceedings of the 1 European Symposium on Particle Classification in Gases and Liquids,* Nuremberg, G4-G14, (1984).
4. **Fraser, S.M., and M.Z. Abdullah**, "LDA Measurement on a Modified Cyclone", in Laser Anemometry, ASME, Fluids Engineering Division, pp 395-403 (1995).
5. **Barth, W**., "Berechnung and Auslegung von Zyklonabscheidern auf Grund neuerer Untersuchungen" *Brennstoff-Warme-Kraft 8*, Heft 1, 1, pp 1-9 (1956).
6. **Zhou, L. X., and S.L. Soo,** "Gas-solid Flow and Collection of Solids in a Cyclone Separator", *Powder Technology*, **63**, pp 45-53 (1990).
7. **Parida, A., and P. Chand**, "Turbulent Swirl with Gas-Solid Flow in Cyclone", *Chemical Engineering Science*, **35**, pp 949-954 (1980).
8. **Boysan, F., Ayers, W.H., and Swithenbank, J.**, "A Fundamental Mathematical Modelling Approach to Cyclone Design", *Trans IchemE*, **60**, pp 222-230 (1982).
9. **Griffiths, W.D., and F. Boysan**, "Computational Fluid Dynamics (CFD) and Empirical Modelling of the Performance of a Number of Cyclone Samplers", *J. of Aerosol Science*, **27** (2), pp 281-304 (1996).
10. **Cristea E. D., E. Malfa, and A. Coghe**, "3-D Numerical Simulation and Measurement of Strongly Swirling Heavy Dust-Laden Flow inside a Cyclone Separator", *Eng. Turbulence Modelling and Experiments*, pp 851-857 (1996).
11. **Hoffmann, A.C., H. Arends, and H. Sie**, "An Experimental Investigation Elucidating the Effect of Solid Loading on Cyclone Performance", Filtration and Separation, pp188-193 (May-June, 1991).
12. **Liden, G., and A.Gudmundsson**, "Semi-Empirical Modelling to Generalise the Dependence of Cyclone Collection Efficiency on Operating Conditions and Cyclone Design", *J. of Aerosol Science*, **28**(5), pp 853-873 (1997).
13. **Hoffmann, A.C., M. de Groot, and A. Hospers,** "The Effect of the Dust Collection System on the Flowpattern and Separation Efficiency of a Gas Cyclone" *Can. J. Chem. Eng.* **74,** pp 464-470 (1996).
14. **Bernard, J.G., J. Andries, and B. Scarlett,** "Cyclone Performance at Various Temperatures", in *Proceedings Filtech Europa '91*, M.J. Taylor, Ed., Karlsruhe, Germany, Oct. 15-17, pp591-599 (1991).
15. FLUENT Inc., Centerra Resource Park, 10 Cavendish Court, Lebanon, NH 03766.
16. **Ray M. B., P. E. Luning, A. C. Hoffmann, M. I. L. Beumer, and A. Plomp**, "Post Cyclone (PoC): An Innovative Way to Reduce the Emission of Fines from Industrial Cyclones", *Ind. Eng. Chem. Res.*, **36**, pp 2766-2744 (1997).

Investigation of Particle/Gas Flow Characteristics in Long Pipelines

B. Golchert, S. L. Chang, and M. Petrick
Argonne National Laboratory, Argonne, IL 60439

There is a growing need by the industry to use computational fluid dynamics (CFD) codes as a tool to improve the operating efficiency or to meet stringent environmental regulations. A successful CFD simulation can provide detailed information on a flow system based on limited experimental data. Using this information, the effects of the operating conditions on the performance of the system can be evaluated to seek optimal operating conditions. Recently, the refinery, chemical, and glass industries identified the CFD capability as a critical area for technological advancement. Many energy conversion processes such as fluid catalytic cracking (FCC) riser reactors, coal-fired combustors, jet engines, and internal combustion engines involve multiphase reacting flows. This type of flow is one of the most difficult to simulate with a CFD code because of the severe numerical problems that often arise out of interfacial interactions with the primary flow hydrodynamics.

Currently, many industrial processes may be improved and their operation better understood by the application of a CFD code. One such process is the conveyance of pulverized coal. The long pipeline of a blast furnace sometimes suffers from particle pile up problems. Argonne National Laboratory (ANL) has used a CFD code to study the particle pile up problem in such a system.

ANL has successfully developed a multiphase reacting flow computer code. This code was developed in several versions for various applications such as FCC riser reactors, coal-fired combustors, and diesel spray combustion. A version of the code is called ICRKFLO which has been validated by comparisons with available experimental data and used to evaluate the performance of several multiphase flow systems. For this particular problem, ICRKFLO solves conservation equations of general flow properties for multiple phases: gaseous species, and solid particles. Phenomenological models were developed to represent major flow processes including turbulence, interfacial drag, and solid-particle interactions. The ICRKFLO code can be applied to general multiphase reacting flows and to non-reacting flows. In this paper, ICRKFLO was used to predict gas/solid flow characteristics in the long pipeline system. The discussion will focus on the prediction of the gas/solid flow characteristics for the pulverized coal transfer lines of a blast furnace. The local gas and solid velocity profiles, the local particle loading, and the local pressure drop are some of the characteristics that are discussed.

Gas-solid flows occur frequently in many industrial processes and systems, such as in pulverized coal transport pipelines and portions of fluid catalytic cracking (FCC) risers. Many of these systems have been modeled and optimized over the years by "engineering know-how," which has been derived from the operating experience with the unit. However, it is being recognized that a more detailed, analytic tool is necessary to improve the efficiency of these systems due to heightened industrial competitiveness. In addition to this competition, increasingly stringent environmental regulations force the industry to improve their systems to comply with these regulations.

In part, these two conditions explain the recent interest in computational fluid dynamics (CFD) applied to industrial systems. A successfully validated CFD simulation can provide detailed information on an industrial flow system based on a limited amount of experimental data. Once a CFD code has been properly validated, it may then be used to suggest optimal operating conditions. Given the wide range of information provided by a CFD code, the term "optimal" may refer to optimal material output, optimal energy efficiency, minimal (optimal) pollutant emission, or optimal lowest operating cost, for example. This versatility has been recently recognized by a number of industries (for example, refinery, chemical, aluminum, and glass), all of which have identified CFD as a critical field for advancement of their technology.

The ultimate goal of applying CFD to the modeling of an industrial system is to produce meaningful data that may be used by the company to improve plant performance. The amount of useable data partially depends on how the system is numerically modeled. Any physical system maybe modeled by a Eulerian or Lagrangian approach (or a mixture of the two for multiphase systems). Many of the recent work in modeling gas-solid flows is based on a Lagrangian model for the solid phase [1-5]. While this approach models the physics of the system on a microscopic level very well, it is often difficult to extract engineering system information from a Lagrangian formulation. This thought has been re-confirmed in a recent paper: "...the particle paths which result from a Lagrangian description often do not contain enough information for engineers" [6]. Therefore, a Eulerian approach was used in this code to model all phases of the system. After validation of the code, it was seen that the Eulerian approach provided information needed to improve a coal transfer system.

NUMERICAL APPROACH

Over the years, the ICRKFLO code was developed at Argonne National Laboratory (ANL) by expanding and enhancing an existing multi-phase reacting flow computer code. The computer code was originally a general CFD code numerically solving conservation equations of mass, momentum, and energy for multi-

Governing Equations

ICRKFLO solves the conservation equations of general flow properties for three phases: gaseous species, liquid droplets, and solid particles. General conservation laws, expressed by elliptic-type partial differential equations, are used in conjunction with rate equations governing the mass, momentum, enthalpy, and species for a three-phase flow with gas species, liquid droplets, and solid particles. For this study, only the solid and gaseous phases are present in a two-dimensional flow. Governing equations include the equation of state and the continuity and the momentum equations. Two additional transport equations are needed for solving the turbulent diffusivity. For convenience in numerical formulation, the governing conservation/transport equations for the gas phase are put in a common form, Equation (1):

$$\frac{\partial}{\partial x}(\theta\rho u\xi - \Gamma_\xi \frac{\partial \xi}{\partial x}) + \frac{\partial}{\partial y}(\theta\rho v\xi - \Gamma_\xi \frac{\partial \xi}{\partial y}) = S_\xi \quad (1)$$

in which ξ is a general gas flow property (scalar, velocity components, turbulent kinetic energy, and turbulent dissipation rate); x and y are spatial coordinates; θ is the gas volume fraction; u, v are velocity components; Γ is an effective diffusivity (calculated from both laminar and turbulent diffusivities); and S_ξ is the sum of source terms applicable to that equation. For all phases, mass and momentum are conserved locally and globally.

The particle phase formulation is derived from an Eulerian approach. In this formulation, the state of the particle-phase flow is governed by the elliptic partial differential equations of fluid mechanics, including conservation of particle number density, and momentum. Particles are generally divided into various groups according to their sizes. Similar to the gas phase formulation, the governing transport equations for the particle phase of the kth size group is put in a common form, Equation (2):

$$\frac{\partial}{\partial x}(n_k u_{c,k}\xi - \Gamma_\xi \frac{\partial n_k \xi}{\partial x}) + \frac{\partial}{\partial y}(n_k v_{c,k}\xi - \Gamma_\xi \frac{\partial n_k \xi}{\partial y}) = S_\xi \quad (2)$$

in which ξ is the general particle flow property (number density, x- and y-momentum); n_k is the particle number density of kth size group; $u_{c,k}$ and $v_{c,k}$ are the particle velocity components of kth size group in the x and y direction, respectively; Γ is the particle diffusivity resulting from interaction with turbulence in the gas phase; and S_ξ is the sum of source terms for each equation. The formulation of the computer code allows for a size spectrum of particles; however, in this study, only one size group of particles was used. The use of mono-sized particles was considered a reasonable first step in utilizing a two-phase flow model to analyze the system.

The momentum equations contain source terms for interphase and intraphase property exchange rates. The computation of these rates may, in some cases, be based on rather elaborate models, which resolve or avoid the numerical instabilities arising out of widely disparate time scales in the macro- and micro-physical scales. Several phenomenological models have been employed to numerically represent these important effects that drive the two-phase flow. The models involve particle-solid interactions, interfacial drag, and multiphase turbulence.

Particle/Particle and Particle/Wall Interaction Models

In a dilute multiphase flow, small particles will generally follow with the gas flow. In a dense multiphase flow, particle-particle and particle-wall interactions become significant and can cause a redistribution of the particles. Experimental data show a U-shaped particle number density distribution in a cross-section of those pipelines with upflow or downflow. The particle number density is an important phenomenon to model properly since it strongly influences the effectiveness of a unit, such as a FCC riser, where the particle number density influences reaction rate and heat transfer. Multiphase hydrodynamic CFD codes have difficulty describing such a U-shape distribution due to the lack of a proper model for particle-solid interactions. A new particle-solid interaction model has been developed. This model accounts for particle-particle and particle-wall collisions in regions of high particle volume fraction. Solid diffusivity and solid stress terms are added to the source terms of the momentum equations. The ICRKFLO code uses this new model in calculations that duplicate the observed U-shape distribution.

In a dense flow region, a moving particle tends to collide with neighboring particles, and the collisions cause the particles to diffuse across the flow stream. Since the particle diffusion process is similar to that of gas molecules, a solid diffusivity can be derived. Since particle collision frequency is proportional to local particle mass flux, the solid diffusivity can be derived as:

$$S_{sd_x} = \mu_{sd} \frac{\partial (n v_s)}{\partial x} \quad (3)$$

The term, Equation (3), is added the particle x-momentum equation to account for the diffusion due to particle-particle collisions. A similar term can be derived for the particle y-momentum equation.

Sometimes, particles are packed in locations where the particle velocity is low. When particles are packed, a particle exerts a force directly on the neighboring particles. A solid stress term is defined for this packed situation. The solid stress is assumed to be proportional to the local solid volume fraction, provided the solid fraction exceeds a preset packed value.

$$S_{sp,x} = \frac{\partial}{\partial x}\left[\kappa(\theta_s - \theta_{sp})\right] \qquad \text{if } \theta_s > \theta_{sp} \tag{4}$$

This term, Equation (4), can be added to the particle momentum equation, and it will only contribute when the particle volume fraction is greater than a pre-set maximum. But, this approach can easily cause numerical divergence because a small change in the calculated solid volume fraction results in a huge change in the solid stress. A better approach, which is used here, is to accumulate local solid stress and calculate solid volume fraction from the solid stress.

Interfacial Drag Model

In a multiphase flow, particles are driven by the gas flow through the drag force.. The drag force is caused by the velocity difference between the gas and solid phases (u_δ, called slip velocity). Empirical drag equations have been commonly used. For a single particle in the gas flow, the drag force of gas exerted on the particle is expressed as a function of the slip velocity kinetic energy, the particle's cross-sectional area, and an empirical drag coefficient. The particle drag force is in the same direction as the slip velocity. The direction of the drag force depends on the sign of the slip velocity. The drag coefficient, C_d, is generally represented by the Reynolds number Re_s.

$$C_d = \frac{24}{Re_s}\frac{(1+0.15Re_s^{0.687})}{1} \qquad \text{and} \qquad Re_s = \frac{2\rho\,|\,u_\delta\,|\,r}{\mu} \tag{5}$$

Not including the collisions between particles, the drag force of the gas on a collection of particles can be obtained by summing up the drag force on the individual particles. Integrating over particle size space for a particle size group, the drag force exerted by the gas on the particles of the ith size lump is shown in Equation (6). This drag force is the source term $S_{u,i}$ of the particle x-momentum equation.

$$S_{u,i} = C_{d,i}\rho\,|\,u_{\delta,i}\,|\,u_{\delta,i}\frac{\pi r_i^2}{2}n_i \tag{6}$$

The interfacial drag forces of the particles and the gas are conserved if the friction dissipation on the particle surface is negligible. The drag force on the gas should be balanced by the drag force on the particles. Therefore, the drag force on the gas can be obtained by summing the particle drag terms over all particle size groups, and it becomes a source term for the gas momentum equation. In a similar way, source terms of the y-direction drag force can be derived for the particle and gas momentum equations.

Multiphase Turbulence Model

A multiphase k-ε model [7] is used in this study. The k-ε model was modified for multiphase flows to account for the turbulent dispersion of particles. The multiphase turbulent transport equations are derived from the single-phase turbulence transport equations by adding a void fraction to the density terms and sink terms in the equations. A sink term is added to the k-equation to account for the loss rate of turbulent kinetic energy in the gas phase as this energy is transferred to particles in the turbulent dispersion process. The loss of the turbulent kinetic energy depends on the particle size. If the particle size is in the sub-micrometer range, the particles can dissipate the turbulence kinetic energy as effectively as the gas,and the loss rate is set to the gas phase kinetic energy convective rate. If the particle size is larger than 100 μm, the loss of turbulent kinetic energy becomes negligible. Particle diffusivity also depends on the particle size. For particles larger than 100 μm, turbulent fluctuations (eddies) have little effect on them and their momentum is only affected by drag computed from the mean gas flow. If the particle size is in a sub-micrometer range, the particles move along with gas flow and have a similar turbulent diffusion effect as the gas phase.

APPLICATION TO A PULVERIZED COAL TRANSFER PIPELINE

This section of the paper presents results from the application of ICRKFLO to a pulverized coal transfer pipeline. These results will give a strong indication of the utility of this code for analyzing and improving two-phase flow systems. The coal transport system between storage bins and a blast furnace consists of a series of long pipes of various lengths, orientation, and flow characteristics. Figure 1 depicts the computational configuration used to model the transfer pipeline. The coal storage bin is located at point A and the blast furnace is located at point H. The points A through H are used to define individual segments of the system.

A majority of the length of this system consists of horizontal flow. It is expected that in these regions there will be some settling of the coal particles (due to gravity), thus creating a stratified flow with coal particles on the bottom and carrier gas on the top. In addition to this settling on the horizontal sections, coal pileup is expected to occur in the elbows where the pileup is caused by the slow response of the particles to the change of flow direction.

It is this change in flow direction in the elbows that creates additional numerical complications when modeling this pipeline. Some of the particles flowing into the elbow will impact upon the curved wall and may adhere or stay near the wall. When a large number of particles begin to reside in the same location, this effect is referred to as pileup. As particles begin to pile up, their presence begins to restrict the effective flow area in the pipe, thus leading to higher velocities and larger pressure drops. The flow patterns inside the elbow will be different depending upon the inlet and outlet piping orientation (horizontal/vertical inlets and outlets).

Once the particles have settled near the wall, they do not numerically "stick" to the wall. Rather, they will stay near the wall until a strong enough force is present to pull them away from the wall. The only forces acting in this system are the force of gravity and the drag force exerted by the moving gas. As was mentioned earlier, the drag force is proportional to the velocity of the gas. Thus, if the gas velocity is low (as it is near the walls of the pipe), the drag force is low. This implies that there is only a weak force trying to pull the particles back into the center regions of the pipe. In reality, this is a transient problem, but because pileup is a rather slow process (on the order of weeks or months for an industrial system), the problem may be treated as steady state.

The flow through these elbows was modeled by ICRKFLO as part of the sectional modeling approach [8]. Several different inlet conditions to the elbows were examined, and the results generated reflect the flow patterns found in industry. The first elbow encountered in the flow system changes downflow into horizontal flow. The particle number density was not as stratified for this first elbow as it will be for later elbows. However, even for this first elbow, Figures 2 indicates that there will be some particle pileup. Since the flow is not very stratified, the pileup is not as severe as that in other elbows. The gas flow field follows the curve of the elbow (results not pictured), while Figure 2 reveals that the particles tend to follow straight lines until they impact on the elbow walls.

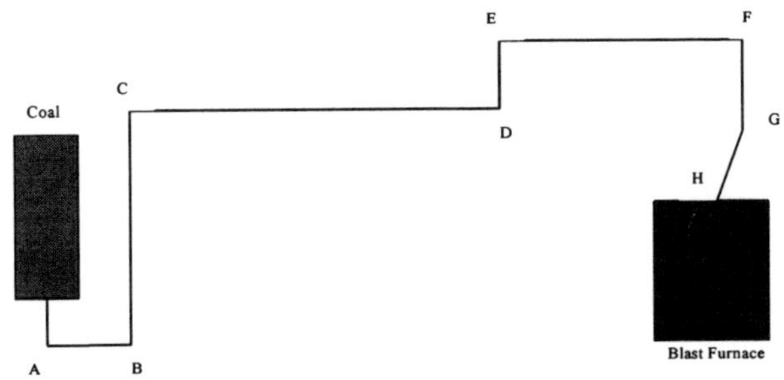

Figure 1: Configuration of a Sample Coal Transport Pipeline

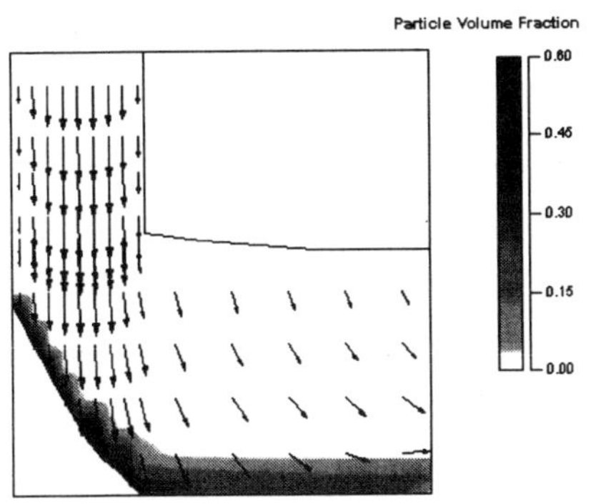

Figure 2: Elbow Turning Down Flow into Horizontal Flow (Particle Velocity Shown)

It is appropriate at this point to note that all of the calculations were done in steady state, even though the actual system is transient. Thus, all of the results from this work are brief "snapshots" of the system. The particle pileup in these figures represents an equilibrium condition. For a converged numerical result, every coal particle that enters the regions of high particle number density kicks out another coal particle back into the flow stream. The total number of particles packed in the system is therefore unchanging.

As was previously mentioned, a CFD code applied to an industrial system should give indications on how to improve the performance of the system. If a CFD code indicates where the particles would pile up in a coal transfer system, then the system designer would know the most likely locations where the pipeline would clog. Figure 3 is a plot of the particle volume fraction for the bottom side of the pipe and for the top side of the pipe. By bottom side, it is meant that radius of the pipe that is farthest "down" in the gravitational direction.

The peaks in Figure 3 represent those locations where the particle packing is high. The black line indicates the bottom surface, and it is on this surface that pileup is expected since gravity will cause the coal particles to settle on the lower surface. The first two peaks on the black line in Figure 3 represent the pileup at elbows A and B. Between these two peaks, the particle volume fraction on the bottom surface is larger than the particle volume fraction on the top surface. This indicates that the particles are settling. The section between the two black peaks is segment BC, a horizontal segment. In this region, the heavier coal particles will settle to the bottom surface because of the pull of gravity. This causes more coal particles to be near the bottom surface with more gas flow near the top surface.

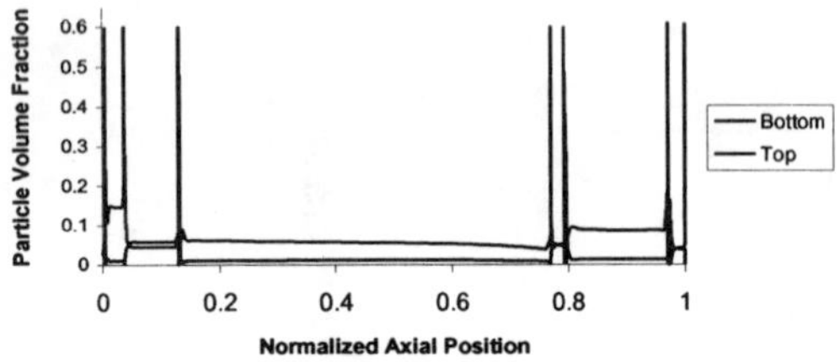

Figure 3: Particle Volume Fraction as a Function of Position along Coal Transport System

Between the second black peak and the first gray peak is segment CD (upflow). With upflow (and downflow), the particle number density distribution (particle volume fraction) is expected to be annular, with both walls of the annulus having approximately the same particle volume fraction. This is evidenced by the black and the gray lines between the two peaks being nearly equal. The bottom particle volume fraction (lower side of the pipe) is slightly higher than the top particle volume fraction due to the increased particle volume fraction on the bottom surface at the entrance to segment CD. The third peak is gray because the coal particles in upflow impinge upon the top wall of the pipe and begin to pile up.

Thus, all of the peaks in Figure 3 may be explained as either particle pileup in an elbow (a black peak) or particle impingement causing pileup (a gray peak). The particle pileup in the elbows represent a serious concern since there is no force present to remove this pileup. The pileup caused by impingement is continually being pulled down by gravity, but the particles are constantly being forced back into the elbow by the impingement of the main flow of particles.

Of all the locations of particle packing, the most serious location is the elbow between segments DE and EF. This elbow occurs at the end of an extremely long horizontal segment. In this segment, the coal particles will settle to the bottom radius of the pipe, and the gas will flow along the top radius of the pipe. This stratification before the elbow will slow down the coal particles. When these slower particles hit the elbow, there is much less force (pressure) present to lift them out of the elbow. Along section DE, more than fifty percent of the total system pressure drop is lost. In addition, segment EF is upflow so the particles must move against the force of gravity. This combination of stratified flow, loss of system pressure, and upflow creates a situation that is difficult to model numerically.

To test the validity of the model, ICRKFLO was set and run for a set of industrial conditions provided by US Steel. The calculated pressure drops were compared with the experimental pressure drop measurements, as can be seen in Figure 4. The agreement between the calculations and the measurements is very good.

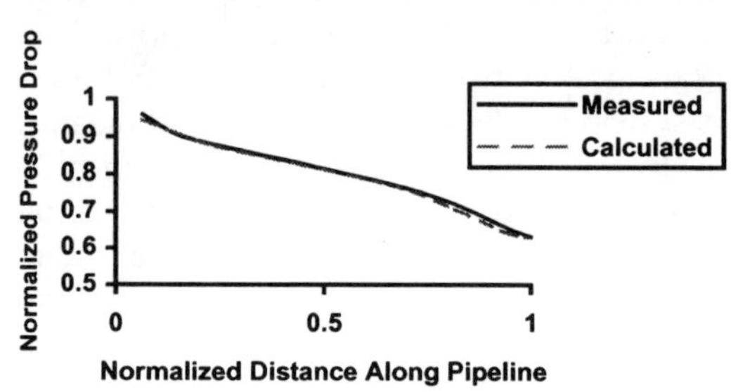

Figure 4: Measured and Calculated Pressure Drop along Pipeline

By using the validated code, a parametric study was performed to determine the effects of operating parameters on the system's performance. The two main parameters that were examined were the coal mass flow rate and the average size of the coal particles. The code shows that with increased particle mass flow rate, the pressure drop for each segment increases. The increased flow rate also increases the pileup in the elbows. Also, the particle volume fraction was found to be affected by the size of the particle. The larger particles tend to pack more readily than the smaller particles, and this phenomenon is also reflected in the computations.

CONCLUSIONS

Improved models have been incorporated into ICRKFLO to better model gas-solid flows. These Eulerian models have been applied to industrial pipeline systems. ICRKFLO correctly reproduces the annual flow expected in upflow and downflow conditions, while the code also properly models the particle settling in horizontal flow. Increased particle mass flow rate and increased average particle size will increase the likelihood of particle pileup. Finally, comparison between

ICRKFLO calculations and system measurements show an exceptionally good agreement, particularly with the pressure measurements and the expected qualitative particle distributions. The results clearly indicate that ICRKFLO is quite capable of modeling these dense two-phase flow systems. Once validated against industrial measurements, the code may then be used as a tool to improve the performance of the gas-solid system.

ACKNOWLEDGEMENT

The authors would like to thank the U.S. Steel Group, a unit of USX Corporation for its support of this work. USX Corporation shall have no responsibility for the validity, originality or applicability of the data contained herein and shall assume no liability for any dissatisfactions or damages resulting from the use of the information presented. The authors would also like to gratefully acknowledge the assistance of Professor Chenn Zhou (Purdue Univ.-Calumet) and Dr. Steven Lottes (ANL).

LITERATURE CITED

1. **Dasgupta, S., R. Jackson, and S. Sundaresan**, "Developing Flow of Gas-Particle Mixtures in Vertical Ducts," *Ind. Eng. Chem. Res.*, **36,** pp. 3375-3390 (1997).
2. **Hrenya, C.M., and J.L. Sinclair**, "Effects of Particle-Phase Turbulence in Gas-Solid Flows," *AIChE Journal*, **43**(4), pp. 853-869 (April 1997).
3. **Lun, C.K.K, and H.S. Liu**, "Numerical Simulation of Dilute Turbulent Gas-Solid Flows in Horizontal Channels," *Int. Journal of Multiphase Flow*, **23**(3), pp. 575-605 (1997).
4. **Shirolkar, J.S., C.F.M. Coimbra, and M.Q. McQuay**, "Fundamental Aspects of Modeling Turbulent Particle Dispersion in Dilute Flows," *Prog. Energy Combustion Sci.*, **22**, pp. 363-399 (1996).
5. **Tashiro, H., and Y. Tomita**, "A Numerical Simulation for Gas-Solid Two-Phase Flow (Influence of the Collision between Particles on the Vertical Upward Flow)," *JSME International Journal, Series II*, **34**(2), pp. 129-133 (1991).
6. **Tu, J.Y., C.A.J. Fletcher, Y.S. Morsi, W. Yang, and M. Behnia**, "Numerical and Experimental Studies of Turbulent Particle-laden Gas Flow in an In-line Tube Bank," *Chemical Engineering Science*, **53**(2), pp. 225-238 (1998).
7. **Zhou, X.Q., and H.H. Chiu**, "Spray Group Combustion Processes in Air Breathing Propulsion Combustors," AIA/SAE/ASME 19th Joint Propulsion Conference, Washington, AIAA-83-1323 (1983).
8. **Chang, S.L., C.Q. Zhou, S.A. Lottes, J. X. Bouillard, and M. Petrick**, "A Sectional Coupling Approach for the Simulation of Multi-Phase Reacting Flow in a Bent Reactor," HTD-Vol. 335, Proc. of the ASME Heat Transfer Division 4:361-373, International Mechanical Engineering Congress and Exposition, Atlanta, GA (November 17-22, 1996).

Saltation Phenomena in the Inlet Regions of Horizontal Pipes

Huseyin Akilli
Department of Mechanical Engineering, Cukurova University, Adana 01330, Turkey
Edward K. Levy
Energy Research Center, Lehigh University, Bethlehem, PA 18015

The gas-particle flow behavior in the inlet region of a horizontal pipe at low conveying air velocities is described. For sufficiently low velocities, particle saltation occurs, allowing some particles to settle onto the lower wall of the pipe and cause buildup of deposits.

Measurements of particle concentration, particle velocity and saltation thickness were performed at various locations along the horizontal pipe using a fiber optic probe. The experiments were carried out m a flow loop with a 0.154 m pipe diameter, air-to-solids mass flow rate ratio in a range of 0.5 to 3.0, and average conveying air velocity ranging from 7.5 to 30 m/s. Pulverized coal with a mean particle diameter of 50μm was used as the solid material. It was observed that conveying air velocity, air-to-solids mass flow rate ratio and inlet condition have a considerable influence on deposit characteristics. Results from numerical simulations indicate particle deposition at conveying air velocities and axial locations in the pipe are in good agreement with the measured values.

INTRODUCTION

The pneumatic conveying of small particles under the influence of gravitational forces is of great interest to mechanical, chemical, civil and environmental engineers. Conveying gas velocity is considered one of the most important factors in pneumatic conveying. There is a substantial economic interest in carrying the solids at low gas velocities, since a lower gas velocity reduces power consumption, pipe erosion and particle attrition. However, in horizontal pipes, below a certain velocity, referred to as the saltation velocity, the particles are unable to remain in suspension and they collect at the bottom wall of the pipe. Deposited particles can significantly reduce the effective pipe cross-section area and can cause pipeline blockages (1). A knowledge of particle deposition is important in the design and reliable operation of pneumatic conveying systems because of its effect on the operating efficiency and safety of the conveying system.

Most of the studies in the literature for horizontal pipes are focused on use of pressure drop measurements to determine saltation velocity. The most-well known correlation for saltation velocity was developed by Matsumoto et al. (2), who observed the saltation velocity of fine particles behaves differently compared to coarse particles. They found that for fine particles, saltation velocity increases as particle diameter decreases, whereas for coarse particles, the saltation velocity increases as particle diameter increases. Tsuji et al. (3) linked this behavior with flow turbulence, reporting fine particles suppress turbulence, whereas coarse particles increase the turbulence level. An average particle size of 350 μm is often used as the cut-off point in determining the distinction between fine particle and coarse particle suspensions (4).

Cook and Hurworth (5) studied the characteristics of deposits as a function of air-to-solids mass flow rate ratio and conveying air velocity. They observed that deposition of particles does not occur immediately after vertical-to-horizontal elbows. The deposit-free length was found to be directly proportional to conveying air velocity. Hoadley (6) reported that deposition first occurs at much lower velocities after a vertical-to-horizontal elbow than after a horizontal-to-horizontal elbow. McCluskey (7) observed the formation of deposits near the elbow exit and concluded that the rope created by the elbow was the main cause of particle deposition.

This paper describes the results of an experimental and numerical investigation of the characteristics of the particle flow in a horizontal pipe at low conveying air

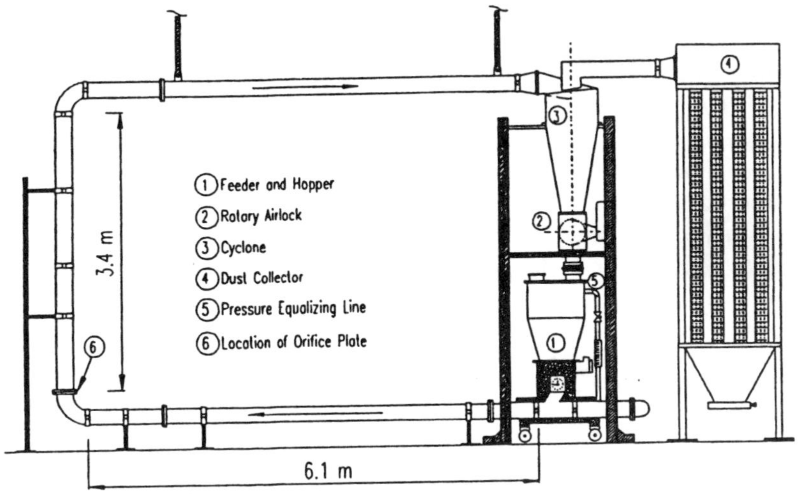

Figure 1. Test facility

velocities. A fiber optic probe was used to measure the effects of air flow velocity, air-to-solids mass flow rate ratio and inlet geometry on particle concentration and velocity profiles and deposit characteristics. In addition, numerical simulations were carried out with a commercial CFD package to determine the effects of conveying air velocity on flow behavior.

FLOW FACILITY

The experiments were carried out in the closed-loop pneumatic conveying system shown in Figure 1. The flow loop consists of two 6.1 μm long horizontal pipes, a 3.4 m long vertical section and two 90° pipe elbows. The saltation measurements were performed in the upper 0.154 m ID horizontal pipe following a 90° vertical-to-horizontal elbow with a radius of curvature-to-diameter ratio of R/D=3.

Conveying air was supplied by screw compressors and air flow rate was measured by an orifice meter. The solids were fed by a calibrated volumetric screw feeder with a variable speed control to adjust solids loading into the conveying line. At the end of the flow loop, the particles were separated from the air stream using a cyclone. The separated particles were fed back to the feeder hopper, permitting continuous operation. Pulverized coal with a mean particle diameter of 50 μm was used as the solid material during the experiments.

Particle concentration (C_p) and velocities (U_p) were measured using a fiber optic probe, which was traversed in the vertical direction from the top of the horizontal pipe downward. The probe contains four glass fibers, two of which are used to transmit the light into the region of the particle-gas flow (8). Light reflected from particles near the probe tip is transmitted back along the other two fibers to photo detectors. The velocity of the particles passing by the probe tip is determined using a cross-correlation technique. Information on local particle concentration is obtained from the mean values of the reflected light signal intensities. Deposit thickness was also measured using the fiber optic probe by noting when the probe indicated high particle concentrations, but with zero particle velocity.

Measurements were performed at ten different instrumentation ports located at the top of the horizontal test section. The distance in the downstream direction from the exit of the elbow is referred to as L. The vertical distance from the bottom wall of the pipe is indicated as x (Figure 2).

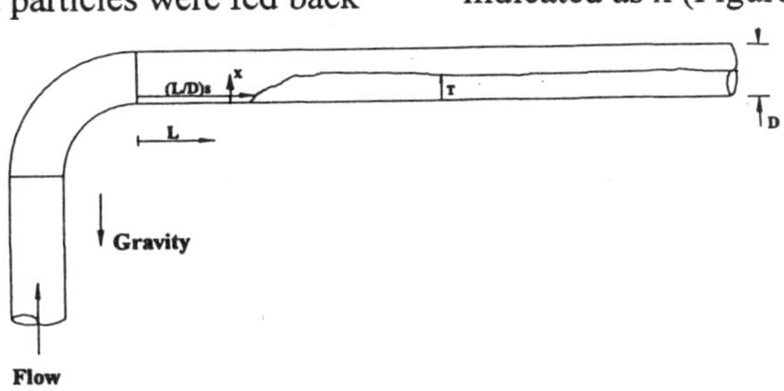

Figure 2. Test section and notation

NUMERICAL MODEL

The numerical simulations were performed with a commercial CFD package, CFX-Flow3D, developed by AEA Industrial Technology. In order to study rope dispersion and particle deposition after a 90° vertical-to-horizontal elbow, the three-dimensional computational model consisted of a vertical pipe with a length of 5 pipe diameters (pipe diameter, D=0.154 m) a 90° vertical-to-horizontal elbow and a horizontal pipe with a length of 30 pipe diameters. Altogether, 56,000 computational cells were used for the simulations.

The flow was assumed to be at steady-state and isothermal for the gas phase. Lagrangian particle tracking along with a two-equation turbulence model (RNG κ-ϵ) was used to model turbulent gas-particle flows through the elbow. Trajectory calculations were performed for 5120 computational particles, each carrying the same mass flow rate. The particles were introduced at 128 starting locations which were selected randomly in the inlet and 40 particles having different particle diameters were tracked at each starting location.

Fully developed turbulent flow was assumed at the inlet of the vertical section. A "no-slip" boundary condition was employed for the gas velocity at the wall. Particle-wall collision was modeled through a coefficient of restitution, which is the ratio of normal velocities of the particle before and after the collision. The coefficient of restitution was set equal to 0.9. The present Lagrangian particle tracking approach does not account for particle-particle interactions.

EXPERIMENTAL AND NUMERICAL RESULTS

Typical vertical variations of particle concentration and velocity along the pipe diameter, obtained from laboratory measurements, are shown in Figure 3 for various locations downstream of the elbow exit (R/D=3, U_{air}=15 m/s, A/F=1). As a result of centrifugal forces in the elbow, particles are forced to the outer wall of the elbow and a highly concentrated particle structure, referred to as a rope, is formed. As seen in Figure 3, the rope is positioned close to the top of the pipe just downstream of the elbow exit plane (L/D=1). Particle concentrations in the rope are typically more than 10 times the average particle concentration in the pipe. Because of momentum losses from the particle-wall and particle-particle collisions in the elbow, the particle velocities in the center of the rope are only one-third of the conveying air velocities. The rope then moves to the bottom wall of the horizontal pipe under the influence of gravity. While the rope is falling to the bottom wall of the pipe, the rope interacts with the high velocity region in the pipe and starts to accelerate and disintegrate. As the rope disintegrates, maximum particle concentrations decrease sharply with axial distance.

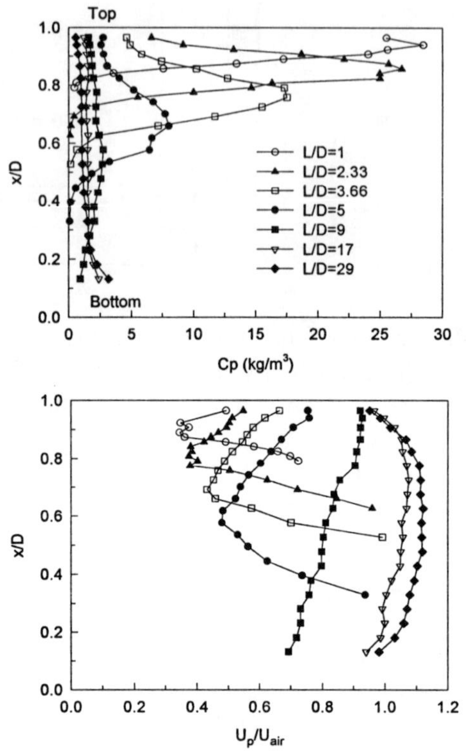

Figure 3: Particle concentration and velocity profiles along the horizontal pipe (R/D=3, U_{air}=15 m/s, A/F=1).

Figure 4 shows typical concentration contours obtained from the numerical simulations. The particle concentration contours at L/D=1 indicate the rope leaves the elbow at the top of the horizontal pipe and secondary flows induced by the elbow carry particles from within the rope around the pipe circumference. The contours at L/D=3.66 and 7 indicate the rope falls to the bottom of the pipe under the influence of the gravity while the rope is dispersing. Most of the particles travel near the bottom wall of the pipe at L/D=29.

Figure 5 illustrates a series of measured particle concentration profiles obtained in the inlet region after the R/D=3 elbow. These data represent three different conveying air velocities (U_{air}=7.5, 15 and 30 m/s) and an air-to-solids mass flow rate ratio A/F=1. The profiles at L/D=1 show a strong particle rope flowing out of the

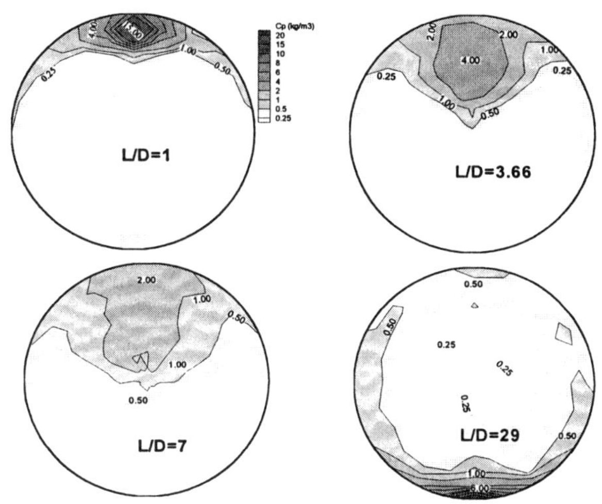

Figure 4: Typical particle concentration contours obtained from numerical simulations.

elbow at the top of the pipe (near x/D=1) with the rope drifting downward towards the bottom wall of the pipe (x/D=0) as the flow moved downstream. Lower particle inertia and smaller turbulent fluctuations caused the rope to drop to the bottom wall of the pipe more rapidly for lower conveying air velocities. At L/D=3.66, the rope was located at x/D=0.88 for a conveying air velocity of U_{air}=30 m/s and at x/D=0.3 for U_{air}=7.5 m/s. The rope dropped to the bottom of the pipe within approximately 4 pipe diameters from the elbow exit plane for U_{air}=7.5 m/s. As the particles then flowed along the bottom wall of the pipe, they were slowed by wall friction and particle-particle interactions causing a stationary particle deposit approximately 5 pipe diameters downstream of the elbow exit plane.

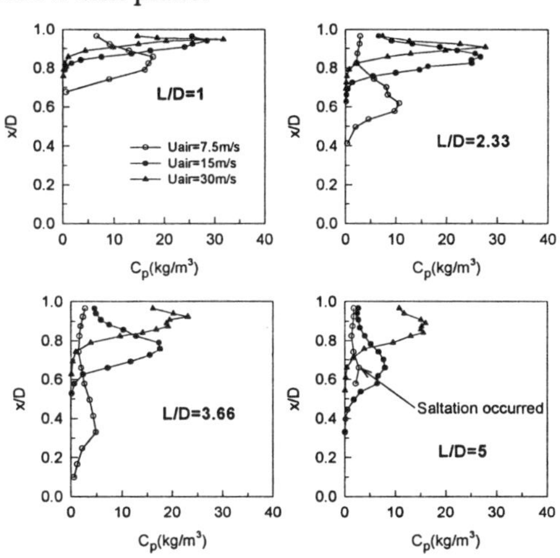

Figure 5: Typical particle concentration profiles after a vertical-to-horizontal elbow, R/D=3, A/F=1. Effects of conveying air velocity.

Figure 6 shows the vertical variations of particle velocity at various locations along the horizontal pipe for the U_{air}=7.5 m/s case. The minimum particle velocity (located within the rope) was approximately one-half of the conveying air velocity at L/D=1. The location of the velocity minimal moved vertically downward as the rope settled towards the bottom of the pipe. As a result of the constriction in the pipe cross-section due to deposit formation, the conveying air velocity and particle velocity both increased sharply after L/D=3.66.

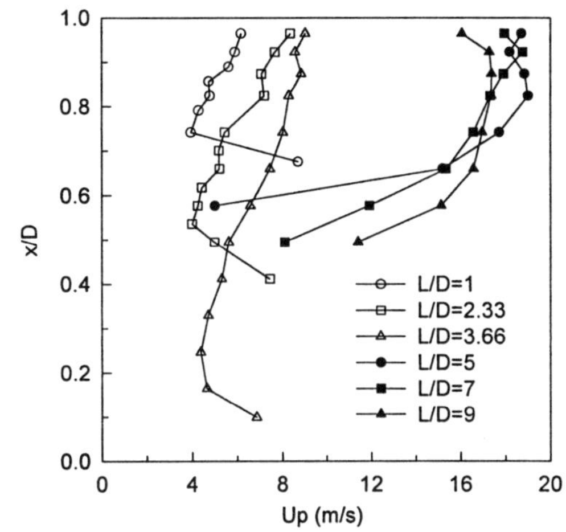

Figure 6: Particle velocity profiles, R/D=3, U_{air}=7.5 m/s, A/F=1.

Figure 7 shows experimental results on the location of the leading edge of the deposit, (L/D)s, as a function of average conveying air velocity and air-to-solids mass flow rate ratio. No particle deposition was observed for conveying air velocities greater than 12 m/s. The leading edge of the deposit moved towards the elbow when conveying air velocity was decreased.

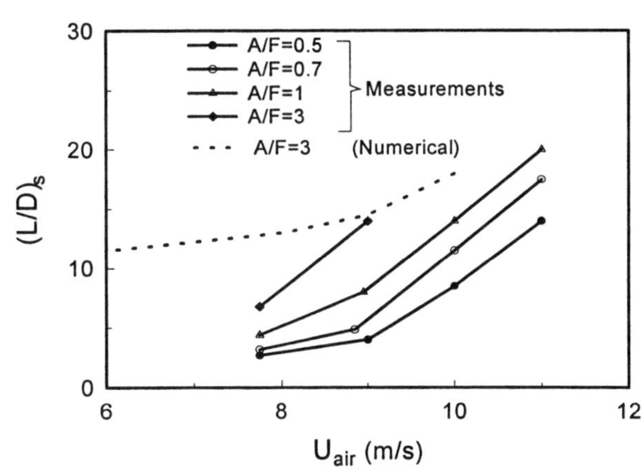

Figure 7: Location of first deposit as a function of conveying air velocity, R/D=3.

In order to investigate inlet condition effects on particle deposition, an orifice plate with an orifice to pipe diameter ratio (β) of 0.7 and a nozzle with β=0.5 were placed at the exit of the vertical-to-horizontal elbow. Figure 8 shows the sketch of the orifice plate, nozzle and plain elbow. For the conveying air velocity of U_{air}=9.5 m/s and air-to-solids mass flow rate ratio of A/F=1, particle deposition occurred at L/D=9 for the plain elbow case (Table 1). The local conveying air velocity and flow turbulence increased due to the constriction of the pipe cross-section when the orifice plate and nozzle were placed at the exit of the elbow. This caused the leading edge of the deposit to move further downstream of the elbow. The deposit was observed at L/D=13 for the orifice case and no deposit occurred in the experimental test section when the nozzle was placed at the exit of the elbow. Similar types of deposit behavior was observed for the R/D=1.5 elbow.

Table 1
The Location of the Leading Edge of the Deposit

	R/D=1.5	R/D=3
Plain Elbow	7	9
Plain Elbow + Orifice (β=0.7)	9	13
Plain Elbow + Nozzle (β=0.5)	17	No deposit

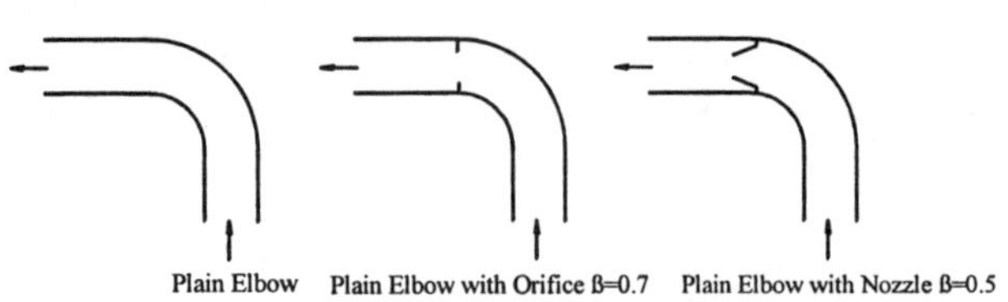

Figure 8: Sketch of devices placed at the exit of the elbow.

Figure 9 shows particle concentration variations obtained from the numerical simulations near the lower wall of the horizontal pipe for different conveying air velocities (R/D=3 and A/F=3). The sharp increase in particle concentration at lower conveying air velocities (6 m/s $\le U_{air} \le$ 9 m/s) is an indication of conditions corresponding to particle deposition at the bottom of the pipe. The simulations also illustrate that as a result of gravity, most of the particles are transported at the bottom wall of the pipe when the conveying air velocity is decreased. The maximum particle concentration reaches approximately 820kg/m^3 for U_{air}=6 m/s at the

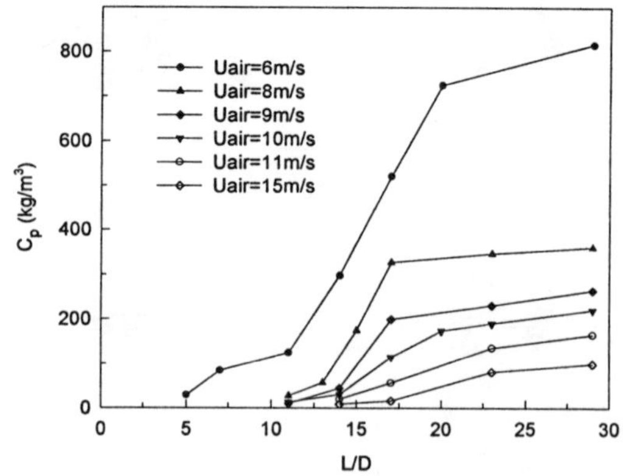

Figure 9: Variation of particle concentration obtained from numerical simulations along the horizontal pipe, R/D=3, A/F=3.

bottom wall of the pipe. This corresponds to a local void fraction in the ϵ=0.5 range, which is typical of a packed bed.

Figure 7 includes numerical results for location of the first deposit as a function of U_{air}. As seen from this figure, the numerical predictions and laboratory data for particle deposition are in good agreement.

CONCLUSIONS

The study dealt with gas-particle flow behavior for low conveying air velocities in the inlet region of a horizontal pipe following a 90° vertical-to-horizontal elbow. The effects of conveying air velocity, air-to-solids mass flow rate ratio and inlet condition on deposition of particles were obtained.

Particle deposits were observed at conveying air velocities lower than 12 m/s in the test section. Deposit thickness and the position of the leading edge of the deposit depend strongly on conveying air velocity. The maximum deposit thickness generally occurred at the leading edge of the deposit. Deposits with irregular axial variations in thickness were observed for higher air-to-solids mass flow rate ratios (A/F>1). Use of an orifice plate or nozzle downstream of the elbow exit delayed the onset of deposition due to higher local conveying air velocities and flow turbulence.

Finally, the numerical simulations show promise for predicting the onset of saltation. Results from the simulations indicate particle deposition at conveying air velocities and axial locations in the pipe which are in good agreement with the measured values.

ACKNOWLEDGMENT

The authors would like to thank the Scientific and Technical Research Council of Turkey for providing financial support to Huseyin Akilli during the course of this study.

LITERATURE CITED

1. **Yan, Y., and Byrne, B.**, *Powder Technology*, Vol. 91, pp. 131-139 (1997).
2. **Matsumoto, S., Kikuta, M., and Maeda, S.**, J. *Chemical Eng. of Japan*, Vol. 10, No. 4, pp. 273-279 (1977).
3. **Tsuji, Y., and Morikawa, Y.**, *J. of Fluid Mechanics*, Vol. 120, pp. 385-409 (1982).
4. **Boothroyd, R.G.**, *Flowing Gas-Solid Suspensions*, Chapman & Hall, London (1971).
5. **Cook, D., and Hurworth, N.R.**, *Pneumotransport 5*, April 16-18, London (1980).
6. **Hoadley, D.**, 1984, *CEGB Report*, TPRD/M/1393N84 (1980).
7. **McCluskey, D.R., Easson, W.J., Greated, G.A., and Glass. D.H.**, *Particle and Particle Systems Characterization Journal*, Vol. 6, pp. 129-132 (1989).
8. **Yilmaz, A. and Levy, E.K.**, "Particle Velocity and Concentration Measurements in Lean Phase Pneumatic Conveying Using a Fiber Optic Probe," FEDSM98-5282, Proceedings of FEDSM'98, 1998 ASME Fluids Engineering Division Meeting, Washington, D.C. (June 21-25, 1998).

Multiphase Opaque Systems: Experiments and Simulations

M.P. Dudukovic
Chemical Reaction Engineering Laboratory (CREL), Washington University, One Brookings Drive, Campus Box 1198, St. Louis MO 63130

Due to space limitations this paper provides only a brief review of the two non invasive techniques for opaque multiphase flows, gamma-ray computed tomography, CT, and computer automated radioactive particle tracking, CARPT, and illustrates their use in our laboratory in studies of bubble columns. Comprehensive treatment of techniques suitable for opaque flows is available in Chaouki et al (1997a, 1997b).

INTRODUCTION

Most multiphase systems of interest in industrial practice (slurry bubble columns, gas-solid risers and fluidized beds, ebullated beds, stirred tanks, solids transport lines) are opaque as they contain a large volume fraction of the dispersed phase. At present, "a priori" predictions of important process parameters such as pressure drop, velocity and holdup profiles, degree of backmixing, etc. is very difficult and relies on correlations which are prone to great uncertainty. Prediction of the needed process parameters based on fundamental fluid dynamic models would be most welcome, yet even the best models require closure forms for phase interaction terms which are still subject to uncertainty and debate. Hence, there is a need to verify such models; verification can only be accomplished if we measure precisely those quantities that we would like the model to ultimately predict, i.e. holdup and holdup profiles, velocity profiles, backmixing, etc. However, the systems are opaque and we cannot "see" into them, and so it seems that a vicious circle has been closed and that model predictions are destined to remain unchecked. Fortunately, as two extensive recent reviews point out (Chaouki et al., 1997a,1997b) there are techniques which can provide us with the desired information. In this presentation we will review two of them: gamma ray computer assisted tomography (CT) for measurement of holdup profiles and computer aided radioactive particle tracking (CARPT) for measurement of velocity profiles and backmixing parameters.

The use of these two techniques is illustrated via the example of bubble columns operated in churn turbulent flow. It is shown that CT can provide time averaged holdup profiles (Kumar et al., 1995), that buoyancy drives the liquid recirculation profile, which can be measured by CARPT (Devanathan, 1991). A simple 1D-momentum balance can relate the observed liquid recirculation with the observed holdup but fails to provide a universal form for eddy viscosity or mixing length or some other closure form. CARPT also provides the full description of large-scale turbulence quantities and a detailed description of the eddy diffusivity tensor (Degaleesan, 1997). All of this has been used successfully to predict liquid backmixing and exit age distribution in both an experimental and industrial column (Degaleesan et al., 1997). Two dimensional axisymmetric CFD simulations based on Euler-Euler two fluid approach are capable of predicting the time averaged liquid recirculation flow field but fail in predicting the holdup profile accurately. Full 3D simulations are needed which illustrate clearly that the nature of the flow does not permit an axisymmetric approximation.

BUBBLE COLUMN EXAMPLE

Due to their importance in gas conversion to fuels and chemicals, as well as extensive use in oxidation, chlorination, hydrogenation and fermentation, I have selected here to deal with measurements of fluid dynamic properties in bubble columns as an example of a multiphase opaque system. A typical bubble column is a cylindrical vessel with gas being sparged into the batch or cocurrent or countercurrent flowing liquid. The direction of liquid flow is not important as long as liquid mass velocity is an order of magnitude smaller than that of the gas. Solids, if present (in slurry bubble columns) and if sufficiently small, follow the liquid and in the first approximation the slurry (liquid + solids) can be treated as pseudo-homogeneous phase. In bubble columns, we have buoyancy driven flows and it is most important to assess the gas holdup distribution in the column that drives the liquid recirculation. Hence, for verification of any CFD model one must have data for gas holdup and liquid velocity distribution in the column as well as data for mixing in the column.

GAMMA RAY COMPUTED TOMOGRAPHY

To measure the gas holdup distribution at any desired cross-section of the column we have implemented a gamma source based fan beam type CT unit. A collimated hard source (100 mCi of Cs-137) is positioned opposite eleven 2" sodium iodide detectors in a fan beam arrangement. The lead collimators in front of the detectors have manufactured slits and the lead assembly can move so as to allow repeated use of the same detectors for additional projections. A 360 degrees scan can be executed at any desired axial location and columns from 1" to 18" in diameter can be scanned in the current configuration. The principle of computed tomography is simple. From the measured attenuation of the beams of radiation through the two phase mixture (projections) we calculate, due to the different attenuation by each phase, the distribution of phases in the cross-section that was scanned. In our CREL CT unit we achieve 3465 to 4000 projections and have a spatial resolution of 5 mm and density resolution of 0.04 g/cm^3. We need a long time to scan the cross-section (about 45 minutes) and, hence, can only achieve the time averaged density distribution. Details of the CT unit are available elsewhere (Kumar et al., 1995; Kumar and Dudukovic, 1997; Kumar et al. 1997). Among the suggested techniques for reconstruction e.g. convolution or filtered back projection, algebraic reconstruction and estimation-maximization algorithms (E-M), we have found the E-M algorithm to be the best (Kumar and Dudukovic, 1997). A sample of the scan for two distributor types shows that at the lower gas velocity of 4 cm/s the sintered plate distributor achieves much more uniform holdup than a perforated plate one, while in churn-turbulent flow at 8 cm/s superficial gas velocity the difference between the distributors is much reduced. Since most of the time averaged holdup distributions are close to axisymmetric, azimuthal averaging provides gas holdup as a function of dimensionless radius. At low gas superficial velocity rather uniform holdup profiles are observed while at high superficial gas velocity, they become almost parabolic.

Typically, time-averaged gas holdup profiles observed in bubble columns can be represented by a power law function of dimensionless radius. The summation of the one dimensional momentum balance for the gas and liquid phase, assuming negligible shear in the gas, provides an equation from which liquid velocity profile is obtainable, provided the gas radial holdup profile is known and either turbulent kinematic viscosity or mixing length or some other simple turbulence closure form is known. Since non-uniform holdup profiles by buoyancy drive liquid recirculation, there is now a need to measure liquid velocities to either test closure forms or develop ideas for improved closures.

COMPUTER AUTOMATED RADIOACTIVE PARTICLE TRACKING (CARPT)

We have adopted CARPT, the technique based on following the motion of a single tracer particle, as a method of mapping the velocity field in the whole system. The technique was introduced for monitoring solids motion in fluidized beds by Lin et al. (1985) and it was adapted for measurement of liquid velocities in bubble columns by Devanathan (1991) and Devanathan et al. (1990). A single radioactive particle dynamically similar to the phase to be traced is introduced into the system. When solids motion is monitored in ebullated beds, fluidized beds or slurries, a particle of the same size and density as the solids in the system is prepared. For monitoring of liquid motion a neutrally buoyant particle is made (Devanathan, 1991). For example, a polypropylene bead of about 2 mm in diameter is hollowed out, a small amount of Scandium 46 is injected, and a polypropylene plug is put in place, so that the density of the composite particle consisting of polypropylene, scandium and air gap equals that of the liquid. Thin film metallic coating assures that bubbles do not preferentially adhere to the particle. An array of scintillation detectors is located around the column. In our case up to 32 NaI 2" detectors are used. The detectors are calibrated in situ with the tracer particle to be used to get the counts-position maps. Conventionally, CARPT calibration was done by positioning the tracer particle (usually Sc-46 of 250 μCi strength was used) at about 1000 known locations and recording the counts obtained at each detector. Various ways of doing this have been described (Devanathan, 1991; Degaleesan, 1997). An alternative way of constructing distance-count maps is via modeling of particle emission of photons and transmission and subsequent detection at the detectors. The Monte Carlo method (Larachi et al., 1994; Gupta, 1998) in which the photon histories are tracked in their flight from the source, through the attenuating medium and their final detection (or lack of it) at the detector can be used for this purpose. Thus, both the geometry and radiation effects are accounted for in estimation of the detector efficiencies in capturing and recording the photons. This involves evaluation of the complex three-dimensional integrals which are evaluated using the Monte Carlo approach by sampling modeled photon histories over many directions of their flight from the source. Once calibration is complete, the tracer particle is let loose in the system and the operating conditions are controlled and kept constant for many hours while the particle is tracked. A regression, least squares method is used to evaluate the position of the particle. Sampling frequency is adjusted to assure good accuracy. Typically, it is selected at 50Hz for bubble columns since the finite size particle used as tracer cannot capture motion of frequencies above 25Hz (Degaleesan, 1997). A wavelet based filtering algorithm (Degaleesan, 1997) is employed to remove the noise in position readings created by the statistical nature of gamma radiation. The processing of data proceeds along the following lines: from filtered particle positions at subsequent times the instantaneous velocity is calculated and assigned to the fictitious column compartment (a compartmental grid is pre-established for the column) into which the midpoint falls. The projection of the particle trajectories in the vertical plane shows that in 100 seconds the particle visits the whole column and in 24 hours of tracking good statistics is obtained everywhere. The run continues until adequate statistics are obtained. For each compartment (cell) ensemble average velocities are evaluated and from the difference between instantaneous and average velocities fluctuating velocity vector is calculated. This allows evaluation of most important Eulerian auto-correlations and cross-correlations from which kinetic turbulent energy and components of the Reynolds-like stresses are obtained (Degaleesan, 1997). Most importantly, the Lagrangian auto-correlations lead to the evaluation of the eddy diffusivities via first-principles (Degaleesan, 1997).

RESULTS FOR BUBBLE COLUMN FLOW FIELD

It is instructive to show some examples of the CARPT results. Even in a churn turbulent flow, at superficial gas velocity of 12 cm/s in a 19 cm bubble column, with a perforated plate distributor, in a time average sense a regular flow pattern of a single liquid recirculation cell is established. If one averages the velocity profiles over the middle part of the column one establishes the typical axial liquid velocity profile which exhibits an inversion point (point where time averaged axial velocity is zero). As gas superficial velocity rises, so does liquid recirculation (Figure 1a). The liquid rises up in the middle and falls down by the walls. The long-

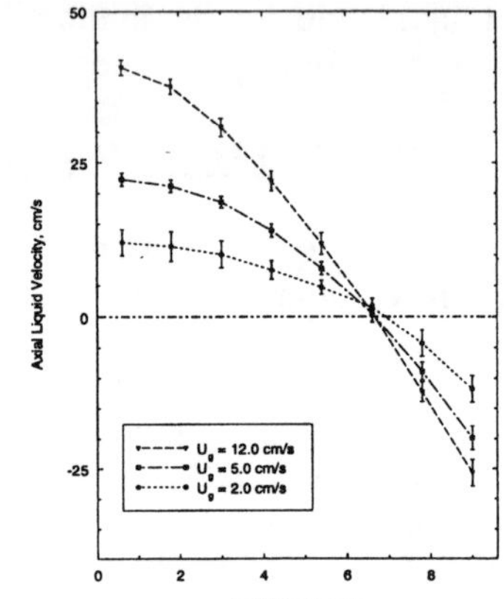

(a)

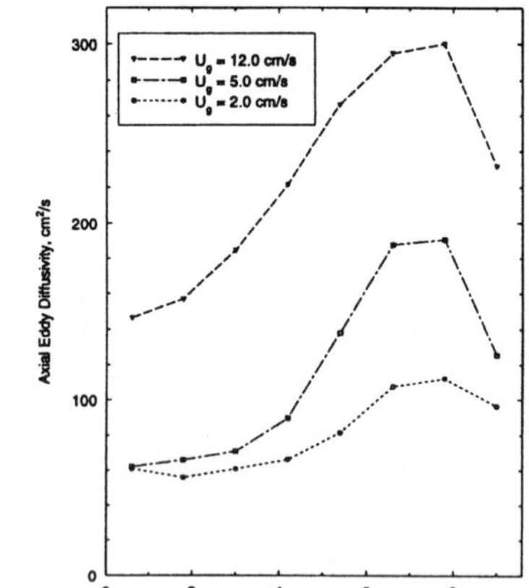

(b)

Figure 1. (a) Time-averaged axial liquid velocity profile in a 19 cm diameter bubble column at three superficial gas velocities. (b)Axial eddy diffusivities in a 19 cm diameter bubble column at three superficial gas velocities

term pattern is axisymmetric. However, at each point (cell, compartment) one can establish a histogram of instantaneous velocities which yield zero means for radial and azimuthal velocities through most of the column. Axial Lagrangian auto-correlation coefficient and axial- radial cross-correlation coefficient can also be established and are indispensable in evaluation of eddy diffusivities (Figure 1b). Turbulent stresses for the 44-cm (18") diameter column in churn-turbulent flow at superficial gas velocity of 10 cm/s exhibit the usual trends for bubble columns. The axial normal stresses are twice as high as the normal radial or azimuthal stresses. The only non-zero shear stress is the r-z one which peaks close to the velocity inversion point. It is important to note that turbulent stresses measured by CARPT are in agreement with the values obtained by Hot Film Anemometry (HFA) and Laser Doppler Anemometry (LDA) under conditions of bubbly flow when the latter two techniques can be made to work (Degaleesan, 1997). This indicates that, at least in buoyancy driven bubble column flows, the major contribution to the $\overline{u'u'}$ and $\overline{u'v'}$ terms comes from relatively large scale low frequency (less than 20 Hz) structures which CARPT can capture and not from the small scale high frequency eddies normally associated with liquid phase turbulent or Reynolds' stresses.

PHENOMENOLOGICAL MODEL DEVELOPMENT

Experimental evidence generated by CARPT-CT can be used to develop a phenomenological model for the column, a model based on experimental observations (Figure 2). Such a model can take advantage of the fact that ensemble averaging of the species conservation equation yields a convection-diffusion equation in which liquid holdup is precisely the long time average measured by CT and liquid velocity is precisely the ensemble average measured by CARPT. Only for the cross-correlation between fluctuating velocity components and fluctuating concentration must one involve a closure. Using Bousinesq hypothesis, and assuming that the diffusivity components are those determined by CARPT, establishes all parameters of the model. Hence, the flow and mixing pattern of the liquid in the column is represented by the liquid long time average recirculation cell with superimposed eddy diffusion. The model can be tested for consistency. In a 8" diameter column with air-water at 10 cm/s superficial gas velocity, liquid holdup profile is established by CT, liquid velocity profile and axial and radial eddy diffusivities are obtained from CARPT. With all of these parameters fed into the convective-diffusion equation the exit age density function can be calculated for an inert tracer. Comparison with an independently conducted tracer experiment shows excellent agreement (Degaleesan, 1997) is indicated in Figure 3. This methodology has then been extended to quantify mixing and flow in an industrial size column for the Advanced Fuels Demonstration Unit (AFDU) in LaPorte, Texas (Degaleesan et al., 1997).

COMPARISON WITH CFD PREDICTIONS

We have qualitatively compared CARPT data (outline of particle trajectory and mean distance traveled) with the Euler-Lagrange simulation of bubbly flows (Devanathan et al., 1994). More than 50,000 bubbles (assuming no bubble-bubble interactions) were simulated in a three dimensional simulation. Reasonable agreement between data and simulations was observed.

Since churn turbulent flow is of most current interest, Lagrange-Euler simulations are not practical, and Euler-Euler interpenetrating two-fluid model presents a natural choice. The issues remain as to how to represent phase-phase interactions such as lift, added mass and drag and how to describe bubble induced turbulence. A simple approach that neglects lift, uses classical approach for added mass, models the drag force based on a sphere affected by presence of other spheres and uses a simple turbulent viscosity concept for bubble induced turbulence (Pan, 1998) has been tried successfully. Comparison of the predictions of the 2D axisymmetric model and data from the 3D cyli[illegible]al column indicate good agreement between

$$\frac{\partial(\varepsilon_L C)}{\partial t}+\frac{\partial}{\partial z}(u_z\varepsilon_L C)=\frac{1}{r}\frac{\partial}{\partial r}\left(r\varepsilon_L D_{rr}\frac{\partial C}{\partial r}\right)+\frac{\partial}{\partial z}\left(\varepsilon_L D_{zz}\frac{\partial C}{\partial z}\right)$$

$$\frac{\partial(\varepsilon_L C)}{\partial t}+\frac{\partial}{\partial z}(u_z\varepsilon_L C)=-\frac{\partial}{\partial z}\left(\varepsilon_L\overline{u_z'C'}\right)-\frac{1}{r}\frac{\partial}{\partial r}\left(r\varepsilon_L\overline{u_r'C'}\right)$$

Convective Diffusion Equation

$$\overline{u_z'C'}=-D_{zr}\frac{\partial C}{\partial r}-D_{zz}\frac{\partial C}{\partial z}\quad;\quad\overline{u_r'C'}=-D_{rr}\frac{\partial C}{\partial r}-D_{rz}\frac{\partial C}{\partial z}$$

Boussinesq Approx.

2-D Convection Diffusion Model for Liquid Mixing

CARPT Experiments indicate D_{zr}, D_{rz} ~ 0

- u_z Ensemble Averaged Liquid Velocity Measured from CARPT
- ε_L Time Averaged Liquid Holdup from CT Measurements
- D_{zz}, D_{rr} Assumed to be CARPT Measured Diffusivities

Figure 2. Two-dimensional convection-diffusion model for the liquid in a bubble

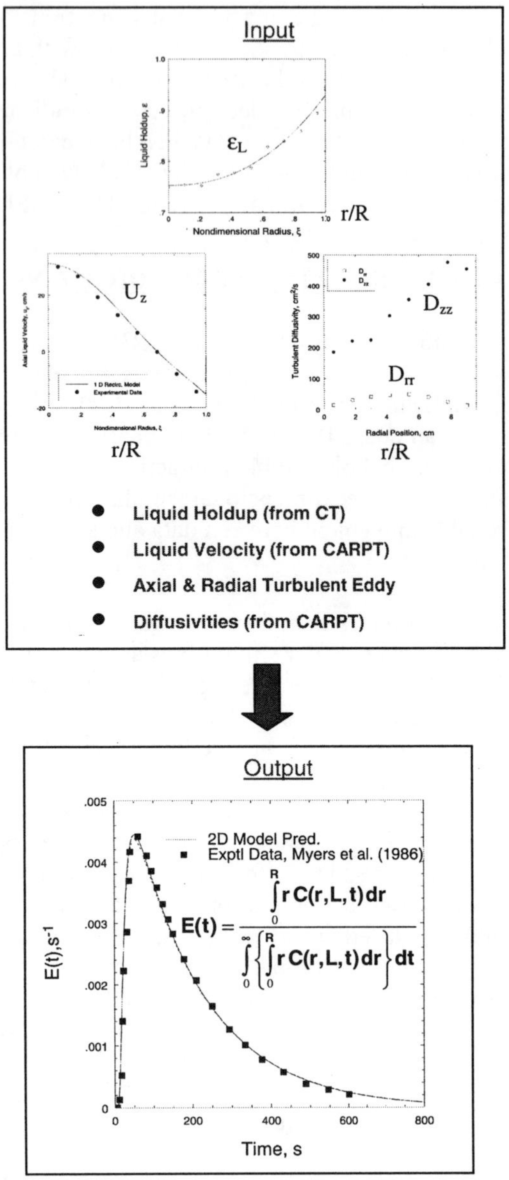

Figure 3 Evaluation of model parameters from CT-CARPT data and prediction of the liquid E- curve

the predicted and measured time averaged velocities (Figure 4). However, the agreement between holdup profiles leaves a lot to be desired. The reason may not lie in the poor choice of closure forms bur rather in the fact that instantaneous flow structures are inherently three-dimensional (3D) and, hence, cannot be properly represented by any transient 2D simulation. A 3D computation (Pan, 1998) reveals the lack of symmetry in the instantaneous velocity patterns and in the holdup and vorticity patterns.

SUMMARY

In summary, it is evident that CARPT-CT are needed in multiphase systems, such as bubble columns, to establish a data base for time averaged as well as statistical properties of the flow over the whole flow domain. This information is then essential in testing the suitability of various CFD models that are challenged with the data. In addition, CARPT-CT offer an

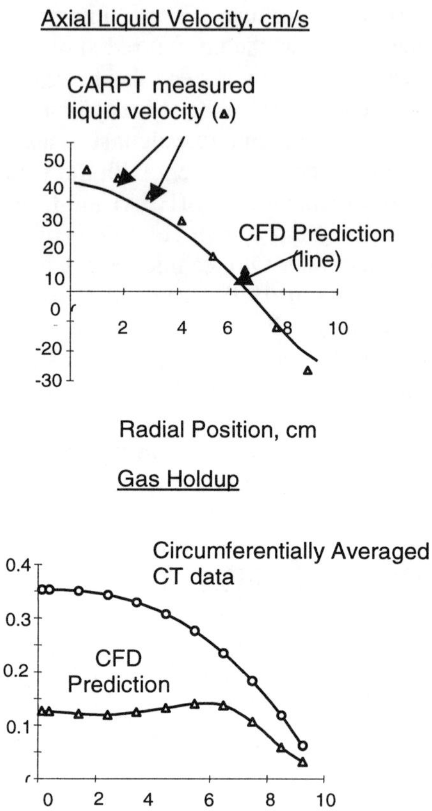

- 3-D experimental data from a cylindrical column (19 cm Dia.)
- 2-D axisymmetric simulation using CFDLIB (Euler-Euler)
- Perforated Plate Distributor (0.1% open area, 0.33 mm holes)
- Superficial Gas Velocity (12 cm/s)
- Mixing Length (1.5 cm), Bubble Dia. (0.5 cm)
- Drag Coefficient, $C_D = 0.44$, Virtual Mass Coefficient, $C_{VM} = 0.5$

Figure 4. Comparison of Euler-Euler two-fluid model predictions for axial liquid velocity and gas holdup profiles, and data from CARPT-CT.

invaluable tool for assessment of the effect of distributor and/or internals on the flow pattern in the column. The data generated by CARPT-CT can be used to develop engineering models of the process equipment studied. An additional use of CARPT was illustrated in fluidized beds by Lyczkowski et al. (1993) where they attempted to relate the experimental information on tracer particle bombardment of heat exchanger surfaces with erosion rates.

ACKNOWLEDGMENTS

Development of CARPT-CT required a team effort since these are not off the shelf techniques. Many individuals contributed extensively to this effort. Among them, the following deserve special recognition: Narsi Devanathan, Yubo Yang, Baisheng Zou, J. M. Mercier, Sailesh Kumar, Sunun Limtrakul, Sujatha Degaleesan, Bente Sannaes, Jinwen Chen, Shantanu Roy, Puneet Gupta and Abdenour Kemoun.

We also gratefully acknowledge the support of our industrial sponsors: Air Products & Chemicals, Chevron, Clariant, DuPont, Eastman Chemicals, Elf, Exxon, ICI, Intevep, Lummus, MEMC, Mitsubishi, Monsanto, Shell, Solutia, Statoil, Union Carbide and UOP, and of the Department of Energy via grant DE FG 95 22 PC 95212 and contract DE FC 95 22 PC 95051.

LITERATURE CITED

1. **Chaouki, J., Larachi, F., and Dudukovic, M.P.** (editors), "Non-Invasive Monitoring of Multiphase Flows", Elsevier, Amsterdam (1997a).
2. **Chaouki, J., Larachi, F., and Dudukovic, M.P.**, "Non-Invasive Tomographic and Velocimetric Monitoring of Multiphase Flows", *Ind. & Eng. Chem. Research* **36**(11), 4476 (1997b).
3. **Degaleesan, S.**, "Fluid Dynamic Measurements and Modeling of Liquid Mixing in Bubble Columns", D.Sc. Thesis, Washington University, St. Louis, MO., USA (1997).
4. **Degaleesan, S., Dudukovic, M.P., Toseland, B.A. and Bhatt, B.L.**, "A Two Compartment Convective Diffusion Model for Slurry Bubble Column Reactors", *Ind. & Eng. Chem. Research* **36**(11), 4670 (1997).
5. **Devanathan, N., Moslemian, D. and Dudukovic, M.P.,** "Flow Mapping in Bubble Columns Using CARPT", *Chem. Eng. Sci.* **45**, 2285 (1990).
6. **Devanathan, N.,** "Investigation of Liquid Hydrodynamics in Bubble Columns via Computer-Automated Radioactive Particle Tracking (CARPT)", D.Sc. Thesis, Washington University, St. Louis, MO., USA (1991).
7. **Devanathan, N., Dudukovic, M.P., Lapin, A. and Lubbert, A.**, "Chaotic Flow in Bubble Columns", *Chem. Eng. Sci.* **50**(16), 2661 (1995).
8. **Gupta, P.**, "Monte Carlo Simulation of NaI Detectors Efficiencies for Radioactive Particle Tracking in Multiphase Flows", *CREL Annual Report*, Washington University, 117 (1998).
9. **Kumar, B.S., Moslemian, D. and Dudukovic, M.P.,** "A γ-ray Tomographic Scanner for Imaging of Void Distribution in Two-Phase Flow Systems", *Flow Meas. Instrum.* **6**(3), 61 (1995).
10. **Kumar, B.S. and Dudukovic, M.P.,** "Computer Assisted Gamma and X-Ray Tomography: Application to Multiphase Flow Systems", Chapter 2 in "Non-Invasive Monitoring of Multiphase Flows", J. Chaouki, F. Larachi and M. P. Dudukovic, eds., Elsevier, Amsterdam (1997).
11. **Kumar, S. B., Moslemian, D. and Dudukovic, M.P.,** "Gas Holdup Measurements in Bubble Columns Using Computed Tomography" *AIChE J.* **43**(6), 1414 (1997).
12. **Larachi, F., Kennedy, G. and Chaouki, J.**, "A γ-ray Detection System for 3-D Particle Tracking in Multiphase Reactors", *Nucl. Instr. & Meth.* **A338**, 568 (1994).
13. **Lin, J.S., Chen, M.M. and Chao, B.T.,** "A Novel Radioactive Particle Tracking Facility for Measurement of Solids Motion in Gas Fluidized Beds", *AIChE J.* **31**, 465 (1985).
14. **Lyczkowski, R. W., Gamwo, I. K., Dobran, F., Ai, Y. H., Chao, B. T., Chen, M. M. and Gidaspow, D.** "Validation of Computed Solids Hydrodynamics and Pressure Oscillations in a Bubbling Atmospheric Fluidized Beds", *Pow. Tech.* **76**, 65 (1993).
15. **Pan, Y.,** "3D Simulation of Bubble Columns", *CREL Internal Report*, (May 1998).

Lattice Boltzmann Simulations of Sedimentation of Cylindrical Particles in a Three-Dimensional Space

Dewei Qi
Department of Paper and Printing Science and Engineering, College of Applied Science and Engineering
Western Michigan University, Kalamazoo, MI 49008

Simulations of sedimentation of cylindrical multi-particles falling against gravity are performed using a lattice Boltzmann method. Calculations of three-dimensional translations and rotations of cylinders are included. The cylindrical particle behavior is dominated by inertia effects associated with wakes. One cylindrical particle may be sucked in strong wakes behind other cylinder to form an inverted "T" cluster. The long bodies of the cylinders turn horizontal dominantly due to a force couple generated by a high pressure at stagnation points. These simulation results are well consistent with the experimental results of three-dimensional cylindrical and disk particles. It is demonstrated that the lattice Boltzmann method can handle cylindrical particles in a three-dimensional space correctly.

1 Introduction

Although there are comprehensive studies on behavior of spherical particle suspensions ([4]; [6]; [5]), very few studies on properties of non-spherical suspensions have been reported. The non-spherical particles in finite-Reynolds number flows are difficult to model due to lack of effective mathematical and simulation methods to handle the interactions between the fluid and the complex geometry of the particles.

The first experimental fluidization of cylinders and disks were reported by Joseph ete. (1987). A narrow channel was used to restrict one layer cylinders to be fluidized under gravity. Three dimensional rotations were allowed for each cylinder in their experiments. They found that the behavior of cylindrical particles is quit different from that of spherical particles. Long bodies of cylindrical particles are dominantly perpendicular to the stream lines of flows. A high pressure at the points of stagnation gives rise to a couple causing the body to turn broadside on. There are very strong wakes behind cylinders. One cylinder may be sucked into the wakes of other cylinder to form an inverted 'T' cluster, which is relatively stable for a while. After the cylinder is sucked into the wake of other cylinder, the strength of the wake may be reduced, and the inverted 'T' cluster may be melt. The formation of inverted 'T' clusters is a feature of cylindrical suspensions. They also observed some doublet and triplet clusters formed by two or more cylinders floating board side on. More information can be found in the article by [8].

In a previous paper [16], the simulations of fluidization of two-dimensional rectangular multi-particles falling against gravity were reported. It was shown that the lattice Boltzmann method has a capacity to deal will non-spherical particles in a two-dimensional space. The same method is extended to deal with three-dimensional cylindrical particles in this paper. Calculatoins of three-dimensional translations and rotations of cylinders are included in this simulation and results will be reported.

2 Lattice Boltzmann Simulation

The lattice Boltzmann method has been developed to simulate non-spherical particles ([10]; [9]; [2]; [12], [13], [14], and [15]) in finite Reynolds number flows. The method in details has been discribed in other article by [15]. I do not repeat it here.

This approach still allows fluid to enter the inside of solid to conserve total mass of fluid at each time step. Conservation of fluid mass guarantees the recovery of the Navier-Stokes equations from LB method. This is particularly important for multiparticles. If the fluid mass in the system is not conserved due to the existence of a single solid particle, then this unconservation of fluid mass will be accumulated due to existence of multiparticles. Therefore, the function of the fluid within solid region is to ensure that the Navier Stokes equations are followed in fluid region.

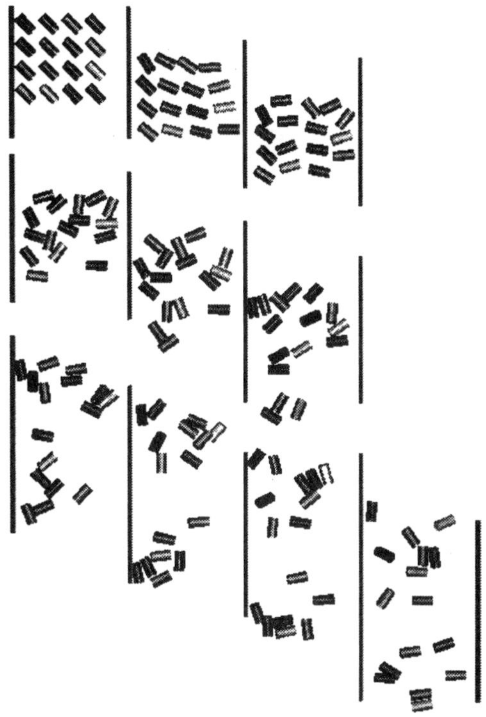

Figure 1: Snap shots of sedimentation of sixteen three-dimensional cylinders with final global $Re = 16.92$. The images in the first row from the left to right correspond to a time order at t=0, 5,000 and 10,000, those in the second row correspond to t= 15,000, 20,000 and 25,000 and those in the third row corresponding to t= 30,000, 35000, 40,000, 45,000.

3 Fluidization of cylindrical particles

3.1 Simulations

The size of the simulation box is 140 × 30 × 150 initially. Four walls are fixed at x=0, x=140, y=0 and y=30, respectively. The length, l, of a cylinder is 24 and the diameter, D, is 12. The ends of the cylinder are rounded by an arc with a radius of 1.5. The cylinder is formed by revolving the rounded rectangular cross section around z-z axis.

The size of the channel used allows the cylinders to rotate in the three-dimensional space. A array of sixteen cylinders are initially located in a regular order in the channel and the Euler angles ϕ, θ, and ψ for all the cylinders are set to 45^0, 90^0 and 0^0, respectively, as shown in the leftmost image in the top row of figure 1. The fluid velocity at the bottom boundary is set to zero and this boundary is always 42.84 ahead the bottom particle. The stress at the top boundary is set to zero and the boundary is always 32.17 behind the top particle. Therefore the simulation box is expanded in the gravity direction during simulation due to relative movement among the particles and the boundary conditions used in this work. The final global particle Reynolds number is 16.92, which is defined by $Re = V_f * l/\nu$, where V_f is the final global velocity of particles. The corresponding solid densities is 1.035. It has been assumed that the fluid density is 1.

Figure 2a shows the global average velocities of all particles in the settling direction. The particle velocities increase during initial settling, then become stable with a small oscillation.

3.2 Orientation of cylinders

The cylinders collectively turn horizontal quickly as shown in the top row of figure 1. Then their positions and orientations become more random due to strong interactions between the cylinders and the flows. However, longer bodies of the cylinders are dominantly in the horizontal direction in an ensemble average. The cylinders in a vertical direction are not stable, they may turn to the horizontal direction. Stable doublets formed by two cylinders floating boardside on are observed, as shown in Figures 1. The polar angle, θ, can be used to characterize the orientation and measure the degree of a cylinder perpendicular to the vertical direction. An angular distribution function (ADF) is adopted to statistically evaluate the orientation of a random suspension of cylinders. The ADF is defined as a probability of finding a single cylinder with a given polar angle θ per unit angle and written as

$$f(\theta) = \frac{1}{n}\sum_i < \delta(\theta) - \delta(\theta_i) > \quad (1)$$

where δ is Dirac delta function, $< ... >$ stands for an ensemble average and n is the total number of the cylinders.

The ensemble average of the angles is carried out for 100 configurations, each 200 time step apart. The results for the periods from t=10,000 to t=30,000 are shown in Figure 2b. It is clearly observed that the long bodies of cylinders broadside on are always dominated during settling as shown in Figure 1 and 2b. The probability is much larger at $\theta = 90^0$ than at other angles, indicating that the horizontal angle ordering is a feature of slender body in sedimenting flows. A high pressure is produced in the front side of a particle and a wake is generated on the back. The high pressure at the

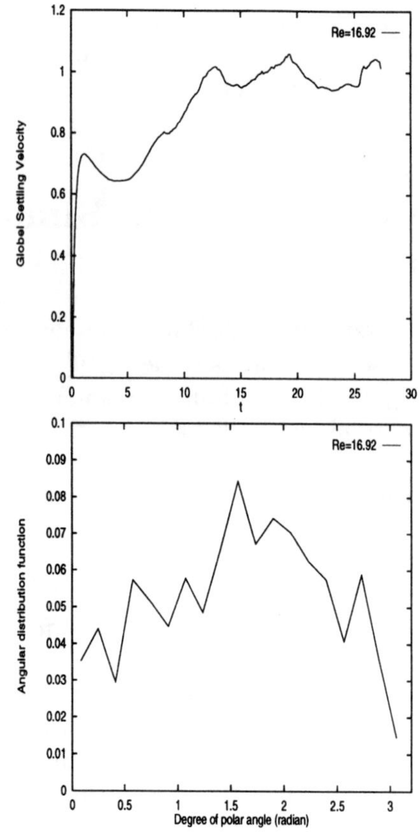

Figure 2: a) The global velocity in the settling direction as a function of time for the cases with final $Re = 16.92$ The velocities have been scaled by global final particle velocity, V_f, and the time has been scaled by D/V_f. b) The angular distribution function (normalized to 1) is shown at final particle Reynolds number $Re = 16.92$. Since the settling direction is in the z-direction, the polar angle $\theta = \pi/2$ corresponds to the horizontal direction.

stagnation points gives rise to a couple causing the body to turn broadside on. This has been explained clearly by [8] and [7].

3.3 Structure of Clusters

One cylindrical particle may be sucked in strong wakes behind other cylinder to form an inverted 'T' cluster, this is a drafting and kissing process. The inverted 'T' cluster may be stable for a short period. After sucking a cylinder, the strength of the wakes may be reduced, and the inverted 'T' cluster is melt or dispersed. The inverted 'T' clusters, each formed by two particles, are shown in Figure 3. In these figures, the inverted 'T' clusters and some doublets are shown, other clusters are intentionally suppressed in order to see them clearly. Drafting, kissing and melting may be used to characterize the behavior of aggregation and dispersion due to multi-particle interaction governed by inertia wake effects. These simulation results are well consistent with the experimental work of [8] who found the similar behavior of non-spherical particles.

A complex velocity field in the simulation box at t=30,000 is shown in figure 4. As expected, there are many wakes behind cylinders.

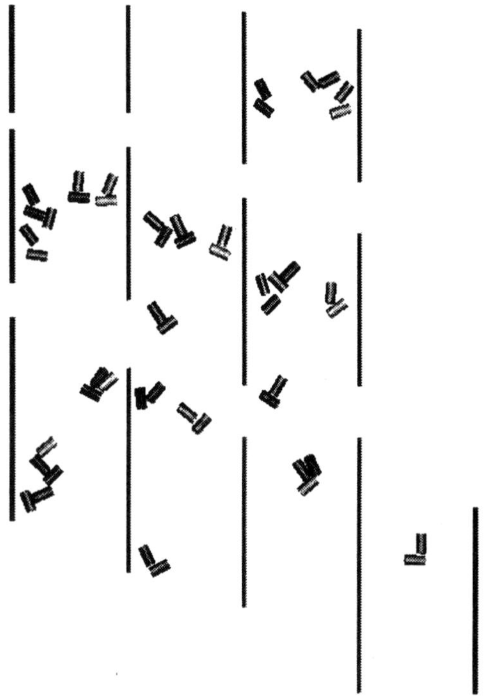

Figure 3: The inverted T clusters formed by two cylinders and some doublets are shown. The configurations are the same as those in figure 1 except that other type clusters are intentionally suppressed to show the inverted T clusters and some doublets more clearly.

4 Conclusions

Fluidization of sixteen cylindrical particles falling under gravity has been simulated by using the lattice Boltzmann method. Calculations of three-dimensional translations and rotations of cylinders are included in the simulation. An angular distribution function is used to describe the orientation of the cylinders and the results show that long bodies of cylinders turn horizontal dominantly during falling. Relative stable doublets formed by two cylinders floating board side on are observed.

One cylinder may be sucked into strong wakes behind other cylinder to form an inverted 'T' cluster. Drafting, kissing and melting may be used to describe the interaction between the cylinders. These behavior dominated by the effects of inertia are well consistent with the experimental works of Joseph. It has been demonstrated that the LB

Figure 4: The projection of velocity on xz plane at $y = H/2$ is shown, where H is the thickness of the simulation box. The section is the same as the rightmost images in the third row of figure 1 corresponding to t=30,000. A complex velocity field with wakes is observed. The shaded circles represent the cross sections of cylinders. The values of the arrows representing the fluid velocities are scaled in an arbitrary unit.

method has a capacity to handle non-spherical particles in the three-dimensional space.

References

[1] AIDUN, C.K. & LU, Y. 1995 Lattice Boltzmann simulation of solid suspensions with impermeable boundaries. *J. Stat. Phys.* **81**, 49.

[2] AIDUN, C.K., LU, Y. & DING, E. 1998 Direct analysis of particulate suspensions with inertia using the discrete Boltzmann equation. *J. Fluid Mech.* **373**, 287.

[3] AIDUN, C.K. & QI, D. 1998 A new method for analysis of the fluid interaction with a deformable membrane. *J. Stat. Phys.* **90**, 145.

[4] BRADY, J.F. & BOSSIS, G. 1988 Stokesian dynamics. *Ann Rev. Fl uid Mech.* **20**, 111.

[5] FENG, J., HU, H.H. & JOSEPH, D.D. 1994a Direct simulation of initial value problems for the motion of solid bodies in a Newtonian fluid. Part 1. sedimentation. *J. Fluid Mech.* **261**, 95.

[6] Happel, J., and R. Pfeffer, " The motion of two spheres following each other in a viscous fluid, A.I.Ch.E. Journal, **6**, 1, 129 (1960).

[7] HUANG, P., FENG, J., JOSEPH, D. 1994 The turning couples on an elliptic particle settling in a vertical channel *J. Fluid Mech.* **271**, 1.

[8] JOSEPH, D.D., FORTES, A.F., LUNDGREEN, T.S. & SINGH, P. 1987 Nonlinear mechanics of fluidization of beds of spheres cylinders and disks in water. in *Advances in Multiphase Flow and Related Problems* (ed G. Papanicolau). SIAM 101.

[9] KOCH, D.L., LADD, A.J.C. 1997 Moderate Reynolds number flows through periodic and random arrays of aligned cylinders. *J. Fluid Mech.* **349**, 31.

[10] LADD, A.J.C. 1994a Numerical simulations of particulate suspensions via a discretized

Boltzmann equation. Part 1. Theoretical foundation *J. Fluid Mech.* **271**, 285

[11] LADD, A.J.C. 1994b Numerical simulations of particulate suspensions via a discretized Boltzmann equation. Part 2. Numerical results. *J. Fluid Mech.* **271**, 311.

[12] QI, D.W. 1997a Non-spheric colloidal suspensions in three-dimensional space. *Int. J. Mod. Phys.* C **80**, 98.

[13] QI, D.W. 1997b Computer simulation of coating suspensions. In *Tappi Fundamental Coating Symposium* 201. Tappi.

[14] QI D.W. 1997c Modeling of fluid transport in an anisotropic fiber network. *APPITA J.* **50**, 319.

[15] QI, D.W. 1999a Lattice Boltzmann Simulations on nonspherical particles in No-zero Reynolds number flows. to appear in *J Fluid Mech.*

[16] QI, D.W. 1999b Lattice Boltzmann Simulations of fluidization of rectangular particles. to appear in *Int. J Multiphase Fluid*

Diameters of Catalyst Clusters in FCC

Weikai Gu
UOP LLC, Des Plaines, IL 60017

The model of gas and clusters is closed by the two constitutive relationships of cluster size and density. Yerushahni and Soong et al. correlated cluster diameters measured in the circulating fluidized bed (CFB) with the room temperature. Hence, the model with the Yerushalmi correlation cannot predict the pressure drop and flow field of commercial fluid catalytic cracking (FCC), because of the high temperature and chemical reactions. This paper gives an analytical expression of the cluster diameters as a function of solid fraction, gas-phase viscosity, and density. Using this constitutive relationship, the gas-cluster model predicted the pressure drops along the risers of several commercial FCC units with reasonable accuracy.

INTRODUCTION

Since 1914, UOP has been developing new concepts, technologies, and designs for refineries. Any new concept has been studied, revised, and refined extensively using pilot plant testing, physical and mathematical modeling and engineering evaluations. In this long development process, UOP needs the capacity to predict a variety of phenomena, for example, the behavior of multiphase flows in a riser, including lift, evaporation, and chemical reaction zones. UOP not only adopted the commercial tools of computational fluid dynamics (CFD) but also developed its own methods for improving the accuracy of its predictions. Before being applied to the commercialization of a new technology, the methods were tested and proved by not only the data from pilot plants and cold-flow modeling but also with commercial data. To guarantee the quality and benefit of a new design, technical and financial resources have been applied to collecting commercial data to benchmark the numerical tools.

Clustering of catalyst particles is one of several major challenges for a numerical tool. An important feature of particle-gas flow in FCC is the tendency for the particles to form aggregate clusters. This phenomenon increases the effective diameter of the interfacial geometry of solid phases up one or two orders of magnitude. As a result, the interfacial drag calculated by the exchange coefficient decreases one or two orders of magnitude. Many of the key features of FCC, for example, the acceleration process of the solid phase, contact-area density between droplets and solid phase, residence time of feed vapor, and total pressure drop over a riser, are generally recognized as being strongly governed by the presence and behavior of clusters. Without a proper description of the clustering process, a numerical model could not give an accurate-enough prediction of hydrodynamic parameters for a new design or revamp. These problems have been attacked by the engineers and scientists at UOP.

Currently, a new RxCat design is proposed for commercialization [1]. This new technology, which mixes spent and regenerated catalyst, almost doubles the catalyst mass flux in risers. To predict riser pressure drop under the riser severe flow conditions imposed by the RxCat system, UOP adopted a gas-cluster flow model developed by Gu and Chen [2]. Because this two-scale model has a new key variable, diameter of clusters, it needs to be closed by the constitutive relationship of cluster size. Yerushalmi and Avidan [3] and Soong et al. [4] correlated the experimental data of cluster diameters measured in the circulating fluidized beds (CFB) with room temperature. However, for commercial FCC, this sort of data is not available for correlation. At this point, only a theoretical analysis can extend Soong's and Yerushalmi's work to commercial units with chemical reaction and heat transfer.

THEORETICAL DESCRIPTION

Strength of Clusters

In CFBs, the moving particles are self-organized into clusters with a certain strength. Like a piece of material, Young's modulus of the solid phase keeps the particle inside the clusters. The normal stress of the solid phase should balance the effective gravity of the cluster. The force balance gives the diameter of clusters as:

$$d_{pc} = \frac{3E_s}{\varepsilon_{pc}\rho_p g(1-\frac{\rho_g}{\rho_p})} \tag{1}$$

Young's modulus is defined as:

$$E_s = (\frac{\partial A}{\partial V})_T = (\frac{\partial U}{\partial V})_T - T(\frac{\partial S}{\partial V})_T$$

where A is Helmholtz function, U the internal energy of particle, and S entropy. Because the internal energy of the solid phase equals zero and the S equals following,

$$S = -\frac{1}{T}\int_{V_{min}}^{V} P_s \, dV$$

the result is

$$E_s = P_s \tag{2}$$

According to the pseudo-thermal theory of granular flow, the random motion of particles causes the pressure or Young's modulus of the solid phase:

$$P_s = \rho_p \varepsilon_{pc}[1 + 2(1+e)\varepsilon_{pc} g_o] T_{pc} \tag{3}$$

where g_o is the radial distribution function deduced by Gidaspow [5]. The granular temperature inside clusters T_{pc} should be solved from the balance of fluctuation energy, $\dot{U}_{pc}$, at the cluster scale because the definition of granular temperature gives:

$$T_{pc} = \frac{2}{3}\dot{U}_{pc} = \dot{v}_{pc}^{\,2} \tag{4}$$

Special Solution of Fluctuation Energy Equation

The equation of fluctuation energy of particles in clusters can be written in the form:

$$\varepsilon_c \rho_c V_c \frac{d\dot{U}_{pc}}{dz} = -\varepsilon_c P_c \frac{dV_{pc}}{dz} + \varepsilon_c \frac{d}{dz}\left(\varepsilon_{pc} k_{pct} \frac{d\dot{U}_{pc}}{dz}\right) - \varepsilon_c \gamma_{pc} + \varepsilon_c \rho_c \phi_{pc} + \varepsilon_c B_{pc} V_c (\dot{V}_{gc}^2 - \dot{V}_{pc}^2) \quad (5)$$

where those terms are convection, work, diffusion, dissipation, source, and interfacial change, respectively. In the cluster scale, the average velocities of all particles in a cluster should be the same. So the first term of the right hand side of the equation vanishes. Once the following two conditions are satisfied,

$$\overline{\dot{V}}_{gc}^2 - \overline{\dot{V}}_{pc}^2 = 0 \quad (6)$$

$$\gamma_{pc} - \rho_c \varphi_{pc} = 0 \quad (7)$$

Equation 5 gives a constant solution for the fluctuation energy of the solid phase in clusters. The first condition means a dynamic balance of fluctuation motions of the two phases. The gas phase does not supply fluctuation energy to the solid phase; it is just a medium for the interaction of particles. In two-phase flows, particles can interact through either collision or interfacial drag. In clusters, particles interact primarily through gas phase, not by collision. The second condition comes from a dynamic balance of fluctuation of the solid phase in clusters. Without collision, the dissipative collision term γ, always equals zero, even for inelastic particles. So the second condition means that the source term of turbulence, ϕ_{pc} , equals zero. In other words, the random motion of particles in a cluster can be self-sustaining. Because of the collisionless interaction of particles, the effective viscosity of the solid phase in two-phase flows should be the viscosity of gas phase. Furthermore, the constant solution of Equation 5 implies that the solid phase has no sheer stress at minimum fluidization conditions. According to the published work, this conclusion has been proved at least by experiments in the bubbling beds [6].

Since the initial condition is minimum fluidization, the result is:

$$\dot{U}_{pc} = \dot{U}_{mf} \quad (8)$$

Minimum Fluidization

From the physical concept of minimum fluidization of the solid phase, the characteristic velocity, length, and frequency of the smallest vortex of the solid phase can be set as:

$$Re_{kmf} = \frac{\dot{v}_{kmf}\, l_{kmf}}{\nu_g} = 1$$

$$Fr_{kmf} = \frac{\dot{v}_{kmf}^2}{g\left(1 - \frac{\rho_g}{\rho_p}\right) l_{kmf}} = 1$$

The smallest fluctuation velocity and vortex length of the solid phase were deduced as:

$$\dot{v}_{kmf} = \left[\nu_g\, g\left(1 - \frac{\rho_g}{\rho_p}\right)\right]^{1/3}$$

$$l_{kmf} = \left[\frac{\nu_g^2}{g\left(1 - \frac{\rho_g}{\rho_p}\right)}\right]^{1/3} \quad (9)$$

$$f_{kmf} = \left[\frac{g^2\left(1 - \frac{\rho_g}{\rho_p}\right)^2}{\nu_g}\right]^{1/3}$$

The fluctuation energy is related with the vortex velocity as follows:

$$T_{pc}=\frac{1}{2}\dot{v}_{kmf}^{2}=\frac{1}{2}[v_g g(1-\frac{\rho_g}{\rho_p})]^{\frac{2}{3}} \quad (10)$$

Combining Equation 2, 3 and 10, the cluster diameter can be written as:

$$d_c(\varepsilon_p=\varepsilon_{plean})\leq$$

$$d_c=C_{dc}\frac{3}{2}[1+2(1+e)\varepsilon_{pc}g_o]\frac{v_g^{2/3}}{[g(1-\rho_g/\rho_p)]^{1/3}}$$

$$\leq d_c(\varepsilon_c=\varepsilon_{mf}) \quad (11)$$

where ε_{pc} is the solid fraction in cluster [2], ε_{plean} the solid fraction of lean phase. The coefficient C_{dc} is used to fit the data, but its order of magnitude should always be one, because the uncertainty of the data should not change the results too much. For Soong's and Yerushalmi's data, C_{dc} equals 0.75. The relative deviation, 25%, may come from the assumption made by Yerushalmi of the constant density in cluster [3].

RESULTS & APPLICATION

Figure 1 shows the curve calculated from Equation 11 and some typical data from Yerushalmi's and Soong's results. The theoretical expression follows the changes in cluster diameter over the whole range of the local solid fraction. It confirms the fundamental physical picture of two-phase flows inside clusters described by Equations. 6, 7, and 10.

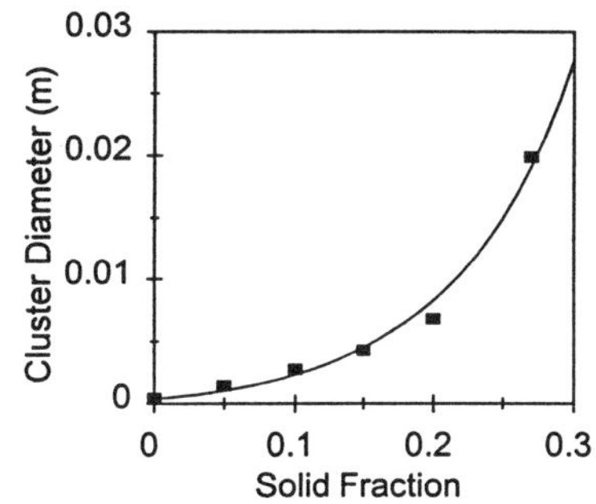

Figure 1. Comparison with Yerushalmi's result of cold flows. Curve was calculated by Equation 11. Dots represent Yerushalmi's result [3].

For commercial units, chemical reactions cause a distribution of the molecular weight of the gas phase. In addition, the temperature in FCC is not uniform because of reactions and heat transfer. The kinematic viscosity of the gas phase in Equation 11 depends on both the molecular weight and temperature of gas phase. Figure 2 demonstrates the effect of gas viscosity on cluster diameters. The Lehigh model used at UOP for the commercial FCC is closed by Equation 11, instead of Soong's and Yerushalmi's correlations [3].

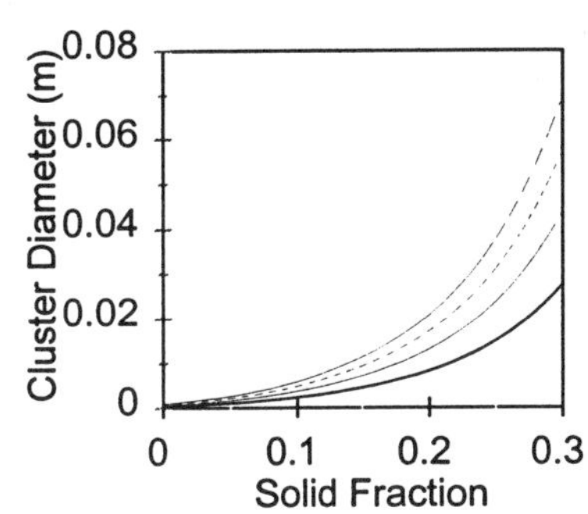

Figure 2. Effect of Gas Viscosity on Cluster Diameter. Solid line for v_g=20mm²/s, dash line for v_g=40mm²/s , doted line for v_g=60mm²/s, thin solid line for v_g=80mm²/s.

This numerical tool is being applied to the new RxCat technology developed by UOP (Figure 3). The new approach used the idea that spent catalyst from a modern-day FCC unit still has useful catalyst activity. The concept, which is based on the reuse of spent catalyst, was formerly known as the X-design. Used, but still active, catalyst is sent to a mixing pot to be blended with regenerated catalyst. This

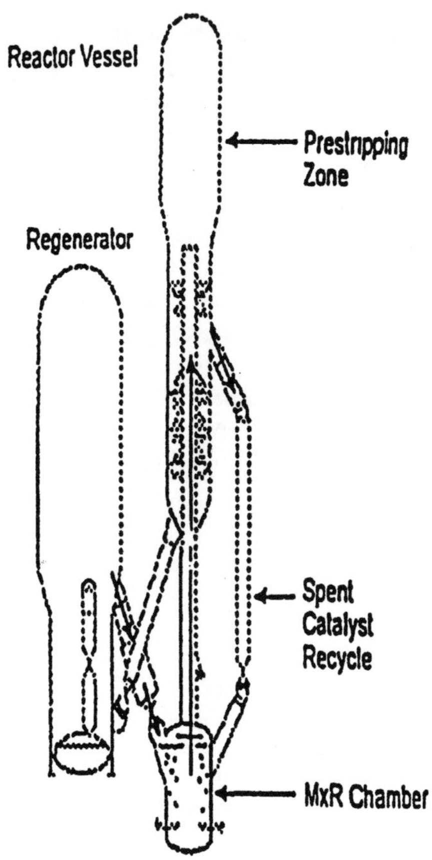

Figure 3. UOP RxCat Configuration

technology can improve the yield and capacity of an existing FCC unit. For the refiner with a FCC unit that is limited by dry-gas production or air blower capacity, a revamp with this technology can increase capacity by up to 8%. For a 30,000 BPD FCC unit, this increased capacity is worth about $8 million per year in increased margin, a reasonable return for a $3 million investment.

To test this numerical tool, UOP measured the pressure drops along the lift zones and reaction zones of commercial units as well as bed-density distributions and chemical samples. These commercial data were used to benchmark the numerical calculations of the model. The results were that the predicted values were within 1.1 to 14 % of commercial data as shown in Table

Table 1. Benchmark with data from 8 units

UNIT	1	2	3	4	5	6	7	8
Number of data	9	4	12	2	1	1	1	3
Average relative error (%)	8	3	11	4	9	1.1	9	8

The total average relative error for the 33 predictions is 8%. Hence, the numerical model with a submodel of clusters is believed to work well. This double-ensemble-averaging model of two-phase flows calibrated by a series of commercial data is being used for the commercialization of the RxCat technology at UOP.

NOTATION

B $\beta/|V_g-V_p|$ for interfacial drag
d diameter
E Young' modulus
e reinstitution coefficient
Fr Froude number
f characteristic frequency
g gravitational acceleration
l characteristic length
P static pressure
Re Reynolds number
T temperature, granular temperature
U internal energy, kinetic energy
V volume, velocity
v characteristic velocity

Superscripts
• fluctuation quantity
__ vector, average quantity

Greek letters
β Exchange or interfacial drag coefficient
γ dissipative collision term
ε volume fraction
ϕ source term of turbulence
ν kinematic viscosity
ρ density

Subscripts
c cluster
g gas phase
gc gas phase in cluster
mf minimum fluidization
kmf condition for the vortex with unit Reynolds and Froude number
p particle
pc particle in cluster
s solid phase
T temperature

LITERATURE CITED

1. **Upson , Lawrence L. and E. Cole Nelson**, "RxCAT Technology for FCC Gasoline," Akzo FCC Symposium, F-10, (June 1998).
2. **Gu , W. and J.C. Chen**, "A Model for Solid Concentration in Circulating Fluidized Beds," *Fluidization IX*, Engineering Foundation, pp.491-499, Durago, Colorado, (May 17-22, 1998).
3. **Yerushalmi, J. and A.A. Avidan**, "High Velocity Fluidization," *Fluidization*, 2nd ed., Academic Press, London, 225, (1985).
4. **Soong ,C.H., K. Tuzla and J.C. Chen**, "Experimental Determination of Cluster Size and Velocity in Circulating Fluidized Bed," *Fluidization VIII*, Engineering Foundation, New York, 219, (1995).
5. **Gidaspow, Dimitri and Lu Huilin**, "Collisional Viscosity of FCC Particles in a CFB," *AIChE Journal*,.**42**.(9), pp2503-2510, (1996).
6. **Clift, Roland**, "An Occamist Review of Fluidized Bed Modeling," *AIChE Symposium Series*, **89** (296), pp1-17, (1993).

Particle Clustering in Down Flow Reactors

S. Krol and H. de Lasa
Chemical Reactor Engineering Centre, Faculty of Engineering Science, University of Western Ontario
London, Ontario, Canada N6G 5B9

It is the goal of the present study to elucidate cluster formation in down flow reactors (d_p=65µm FCC particles). Modeling is based on particle mechanics, with particle evolution being pictured through the dynamic interaction of trailing and leading particles. With this goal, the equations of motion are solved. It is demonstrated that the hydrodynamic particle interaction leads to the formation of agglomerates (clusters). Modeling also allows to calculate the particle approaching times and to postulate a cluster formation mechanism with agglomeration being the result of trailing particles "catching" leading particles from the back.

A characteristic property of solids particles suspended in circulating and turbulent fluid beds is the existence of two regions: a core and an annular region [1,2].There is in these fluidized beds ample opportunity for particle clustering with the formed clusters evolving in a continuous gas phase.

Regarding downer units, there is also suggestion in the technical literature, that particles can agglomerate in down flow reactors [3,4]. Agglomeration is invoked to explain measured slip velocities several times the terminal velocity of individual particles [3,5] and this for superficial gas velocities, u_g>4 m/s [6]. There is also the implication that slip velocities can be affected by the wide particle size distribution of the FCC catalyst [7].

In the present study, it is demonstrated that cluster formation in downer units is the result of a specific mechanism with particles forming "string" type of agglomerates.

Particle Clustering. Modeling and Theoretical Considerations

As already stated, a significant issue for the novel down flow reactors is the way particles evolve in the unit, either as single particles or clusters/strings of particles. The goal here is to consider a model where particles agglomerate on the basis of leading particles being followed by trailing particles.

Correspondence concerning this paper should be addressed to H. De Lasa.

For instance, a trailing particle experience a smaller drag at some point while approaching a leading particle and the two particles become in contact forming a string of two. The process may repeat as the string of two, becomes now the trailing particles, approaching another particle and thus, forming a string of three. This process continues until the string becomes unstable, collapse and the process restart again.

The basic phenomena occurring in a down flow particle suspension and favoring particle clustering is the result of various forces acting on moving particles:

$$F = \sum_{1}^{4} F_k \tag{1}$$

being F_1 or $F_{a,i}$ the added mass force, F_2 or $F_{d,i}$ the drag force, F_3 of $F_{g,i}$ the gravity/buoyancy force and F_4 or $F_{B,i}$ the Basset history force.

Regarding non-hydrodynamic forces, they can be of repulsive-attractive type (electrostatic forces) or attractive type only (adhesive forces). Under expected conditions in down flow reactors attractive forces prevail and this may yield particle agglomerates of slightly larger sizes than the ones observed neglecting non-hydrodynamic forces. Thus, an analysis excluding external forces, as the one attempted here, provides conservative estimates of particle agglomeration.

Then, taking into account eq(1) and

various model considerations [8], it is proposed for both trailing and leading particles the following set of motion dimensionless equations:

$$\frac{d\bar{v}_{s,L}}{d\bar{t}} = \lambda_L(1-\beta)\bar{g} - \frac{3}{4}\frac{\lambda_L\gamma_L\beta d^2_{p,L}\tau v_{t,L}}{d^3_{v,L}}\bar{v}^2_{s,L}C_{d,L} + \frac{6\lambda_L\tau}{\pi d^3_{v,L}\rho_p v_{t,L}}F_{B,L} \quad (2)$$

and

$$\frac{d\bar{v}_{s,T}}{d\bar{t}} = \lambda_T(1-\beta)\bar{g} - \frac{3}{4}\frac{\lambda_T\gamma_T\beta d^2_{p,T}\tau v_{t,L}}{d^3_{v,T}}\bar{v}^2_{s,T}C_{d,T} + \frac{6\lambda_T\tau}{\pi d^3_{v,T}\rho_p v_{t,T}}F_{B,T} \quad (3)$$

with $C_{d,L}$ and $C_{d,T}$ being the drag coefficients defined as reported in Table 1.

Reviewing eqs (2) and (3) it can be observed that these equations can be expressed in dimensionless terms using the terminal velocity of a single leading particle ($v_{t,L}$), a characteristic time ($\tau = d_{p,L}^2\,\rho_p/18\mu$) and a distance ($d_{p,L}$).

Eqs(2) and (3) together with,

$$\frac{d\bar{D}}{d\bar{t}} = \frac{\tau v_{t,L}}{d_{p,L}}[\bar{v}_{s,L}(\bar{t}) - \bar{v}_{s,T}(\bar{t})] \quad (4)$$

determine the change of the dimensionless distance ($\bar{D}$) with the dimensionless time ($\bar{t}$) or in other words the time needed for the trailing particles to reach the leading particles (t_{ap}= approaching time). Note that in the present study the calculated approaching times are estimated with steady gas superficial velocities.

Initial conditions for this set of equations are the following:

$$\bar{v}_{s,L}(0) = \bar{v}^o_L;\ \bar{v}_{s,T}(0) = \bar{v}^o_T;\ \bar{D}(0) = \bar{D}_o \quad (5)$$

with $\bar{D}_o = 5/2(1 + d_{p,T}/d_{p,L})$

Simultaneous numerical solution of eqs(2),(3) and (4), provides the results reported in Figures 1 to 3. An important aspect considered in these figures is the effect of initial velocities (refer to eq(5)) on the clustering phenomena . Note that the initial condition for this calculation assumes that both the leading and the trailing particles can have initial velocities varying in a range between zero and several times the terminal velocity. Thus, an analysis like the one developed demonstrates that wide variation of particle velocities, such as the one expected in the downer feeding section, may strongly affect clustering phenomena.

Modeling Results

Fig.1 reports modeling results for the case of one trailing particle evolving at terminal velocity and a leading cluster containing one, two, three and four particles. Note that the leading cluster is hypothesized to assume different initial velocities. It is shown that the resulting approaching times (Fig.1) are influenced by both the initial leading particle velocity ($v_L^o/v_{t,L}$) as well as the number of particles involved in the leading cluster. Even more it can be observed that approaching times only remain at reasonable levels, let say below 30 milliseconds, if the leading cluster initial velocity is below a maximum initial critical velocity(broken vertical line in Fig.1) . If the initial leading cluster velocity is larger than this critical velocity agglomeration does not occur. On the other hand if the initial velocity is smaller than this critical value, particle clustering takes place. Thus, the cases considered in Fig.1 demonstrate that there is limited opportunity for cluster formation through a mechanism where a leading cluster is caught by a single trailing particle arriving from the back.

On the other hand, Fig.2 reports modeling results for the case of one leading particle evolving at terminal velocity and a trailing cluster constituted by one, two, three and four particles. Note that different initial velocities ($v_T^o/v_{t,L}$) were assigned to the trailing cluster. Fig.2 demonstrates that approaching times are quite affected by both the initial velocity as well as the number of particles involved in the trailing cluster. It can, however, be observed that approaching times remain in almost all cases at reasonable levels with the only exception of the case of "one" leading cluster interacting with "one" trailing particle. In any case, these results strongly suggest that agglomerate formation is mainly due to trailing particles meeting a leading particle and catching it from the back.

Regarding the simulations of Figs. 1 and 2 they were developed assuming that the particles forming the leading and the trailing clusters have 65μm size. In practice, it is well known that particles case of FCC catalyst have a particle size distribution. Thus, in order to illustrate the effect of particle size distribution on the above described phenomena the following cases were considered (Fig.3.): a)45μm trailing particle and 85 μm leading particle; b) 65μm trailing particle and 65 μm leading particle; c) 85μm trailing particle and 45 μm leading particle. It can be observed that the increase of the size of the leading particle versus the trailing particle (case a)) gives a reduced critical initial velocity (broken vertical line). On the other hand, the increase of size of the trailing particle versus the leading one gives an increased critical initial velocity. On this basis, it can be concluded that particle size distribution while changing somewhat the interaction between leading-trailing particles does not change the expected overall mechanism where clusters are going to be formed predominantly as a result of trailing particles catching leading particles from the back.

While no specific argument is advanced in this study about cluster stability it is expected that clusters do no grow indefinitely and clusters may collapse at same point. Let say a cluster of 5 particles could break in fragments of 3 and 2, and will be, following the principles of particle mechanics, reformed eventually in few milliseconds.

Conclusions

a) Modeling based on particle mechanics, pictured through the interactions of trailing and leading particles, allows to calculate the particle approaching times and to show the significance of particle clustering in down flow units.
b) These strings of particles are subject to the laws of cluster mechanics with trailing particles "catching" leading particles form the back being the dominant particle agglomeration mechanism.

Notation

a_i acceleration coefficient = $(d_{p,i}/v_i^2)dv_i/dt$ (m s^{-2}).
A_i' particle projected area (cm^2).
$C_{d,i}$ drag coefficient for either the leading or the trailing particles including hydrodynamic interactions(-).
$C_{d,i}^o$ drag coefficient for single leading or the trailing particles calculated after Clift et al.[13].
$d_{p,i}$ diameter of either the leading or the trailing particle (cm).
D distance between centers of leading and trailing particles (cm).
$\overline{D}$ dimensionless distance between particles $=D/d_{p,i}$(-).
$\overline{D}_o$ dimensionless initial separation between centers of leading and trailing particles(-)
F net force (refer to eq(1)) (newtons).
$F_{a,i}$ added mass force for either the leading or the trailing particles (newtons).
$F_{B,i}$ Basset history forces for either the leading or the trailing particles (refer to [8]) (newtons).
$F_{d,i}$ drag forces applied on either the leading or the trailing particles(refer to [8]) (newtons).
$F_{g,i}$ gravity/buoyancy forces (newtons).
g gravity acceleration (m s^{-2}).
$\overline{g}$ dimensionless gravity acceleration = $g\tau/v_{t,L}$ (-).
Re_L Reynolds number of leading particle based on volume equivalent diameter (-).
$\overline{t}$ dimensionless time (-).
t_{ap} approaching time. Time needed for the tracking particle to reach the leading particle (ms).
$\overline{v}_i^o$ dimensionless velocity of either leading or trailing particle at zero time. The terminal velocity of single leading particle($v_{t,L}$) taken as reference (-).
v_L velocity of leading particles (m s^{-1}).
$\overline{v}_L$ dimensionless velocity of leading particles (-).
$\overline{v}_{s,i}$ dimensionless slip velocity =[cluster velocity-superficial gas velocity]/$v_{t,L}$ (-).
v_T velocity of trailing particles (m s^{-1}).
$\overline{v}_T$ dimensionless velocity of trailing particles (-).
$v_{t,L}$ terminal velocity of leading particles (m s^{-1}).
$v_{t,T}$ terminal velocity of trailing particles (m s^{-1}).

Greek Symbols

α_i coefficient of added mass with subscript "i" representing the leading or trailing particles (-).
β ρ_g/ρ_p (-).
γ_i $A_i'/[\pi d_{p,i}^2]$, dimensionless group(-).
λ_i $1/[1+0.5\alpha_i\beta]$, dimensionless group with α_i defined as reported in Table 1(-)
ρ_g density of the gas (kg m^{-3}).
ρ_p density of the particle (kg m^{-3}).
μ gas viscosity (kg m^{-1} s^{-1}).
τ $\rho_p d_{p,L}^2/18\mu$ (s).

Superscripts

o refers to a time, t=0
i refer to L (leading) or T (trailing) particles

Literature Cited

[1] Ege P., Grislingas A., de Lasa H., **Chem. Engng. J.** 1996, 61, 179-190
[2] Farag H., Ege P., Grislingas A., de Lasa H., **Can.J.Chem.Engng.** 1997, 75, 851-860
[3] Zhu J.-X., Yu Z.-Q., Jin Y., Grace J.R., Issangya A., **Can. J. Chem. Engng.**, 1995, 73, 662-677.
[4] Bai H., Jin Y., Yu Z., Gan N. in Circulating Fluidized Bed Technology III, Basu, P. (ed.), Pergamon Press, 1991, 157-163.
[5] Yang Y.-L., Jin Y., Yu Z.-Q., Zhu J.-X., Bi H.-T., **AIChE Symp. Series**, (1993) No 296, vol.89, 81-90.
[6] Aubert E., Barreteau D., Gauthier T., Pontier R., in CFBT IV, Basu, P. (eds.), Pergamon Press, 991, 403-405.

[7] Herbert P.M., Gauthier T.A., Briens C.L., Bergougnou M.A. , **Powder Tech.**, 80, (1994) 243-252.

[8] Krol S., Pekediz A., de Lasa H., Submitted to **Powder Tech.**, March (1998)

[9] Tsuji Y., Morikawa Y., Shiomi H., **J. of Fluid Mechanics**, 139 (1984) 417-434.

[10] Vojir D.J., Michaelidis E.E., **Int. J. Multiphase Flow**, 20 (1994), 547-556.

[11] Taniguchi I., Asano K., **J. Chem. Engng. of Japan**, 20 (1987) 287-294.

[12] Taniguchi I., Asano K., **J.Chem.Engng of Japan**, 22 (1989) 701-703.41].

[13] Clift R.,Grace J.,Weber M.,Bubbles, Drops and Particles, Academic Press,New York (1978).

Table 1. Model Parameters

Parameter	Ref.
$\alpha_i=[0.066\, a_i^2/(1+0.12\, a_i^2)]$ with $a_i=(d_{p,i}/v_i^2)\, dv_i/dt$	[9] [10]
$C_{d,L} = \dfrac{C^o_{d,L}}{[1+0.32\,(\frac{d_{p,T}}{d_{p,L}})^{2.22}\,(\frac{D}{d_{p,L}})^{-1.4}\, Re_L^{0.42}]}$ $C_{d,T} = \dfrac{C^o_{d,T}}{[1+0.53\, Re_L^{0.31}\,(\frac{d_{p,T}}{d_{p,L}})^{-0.43}\,(\frac{D}{d_{p,L}})^{-0.9\, Re_L^{-0.06}}]}$ with $C_{d,L}^o$ and $C_{d,T}^o$ being the drag coefficients for isolated leading and trailing particles respectively.	[11] [12]

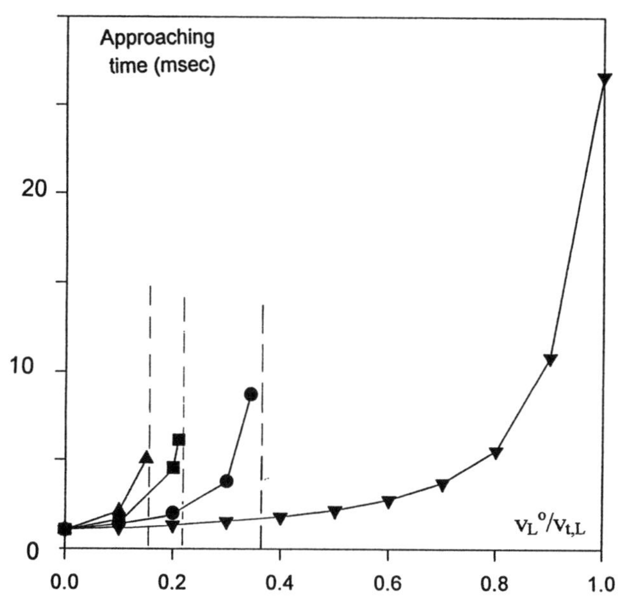

Fig.1. Change of approaching times with $v_L^o/v_{t,L}$
(▼) 1 leading particle, 1 trailing particle,(●) 2 leading particles, 1 trailing particle
(■) 3 leading particles, 1 trailing particle,(▲) 4 leading particles, 1 trailing particle

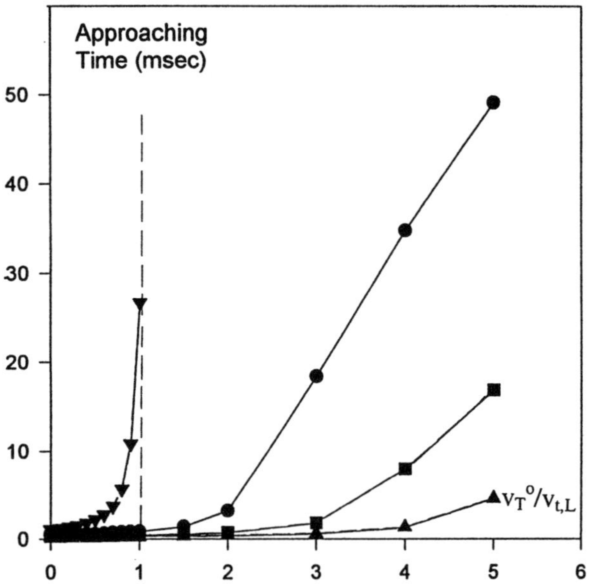

Fig.2. Change of approaching times with $v_T^o/v_{t,L}$
(▼) 1 leading particle, 1 trailing particle,(●) 1 leading particle, 2 trailing particles
(■) 1 leading particle 3 trailing particles,(▲) 1 leading particle, 4 trailing particles

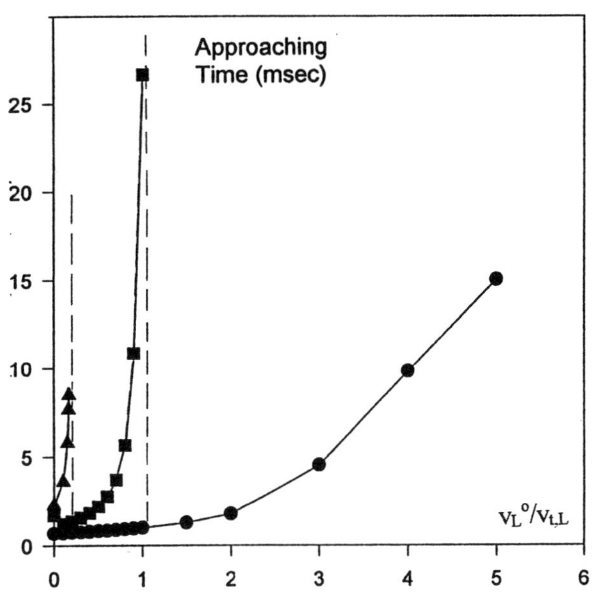

Fig.3. Change of approaching times with $v_L^o/v_{t,L}$
(▲) 45 μm trailing particle and 85 μm leading particle.
(■) 65 μm trailing particle and 65 μm leading particle.
(●) 85 μm trailing particle and 45 μm leading particle.

Describing Mass Transfer in Circulating Fluidized Beds by Ozone Decomposition

Olav Bolland
Norwegian University of Science and Technology, Department of Thermal Energy and Hydropower, N-7034 Trondheim, Norway
Rainer Nicolai
Institute of Process Engineering, Swiss Federal Institute of Technology, ETH-Zentrum ML G16 8092, Zurich, Switzerland

The objective of this work was to improve understanding of the mass transfer between gas and solids in the riser section of gas-solids circulating fluidized beds (CFB). In order to experimentally investigate gas-solids mass transfer in a CFB, a suitable low-temperature mass transfer process was applied: the decomposition of ozone (0_3) using particles as catalyst. Particles used were angular cast steel with an average diameter (mass mean) of 134μm. The experiments were carried out in a semi-industrial sized CFB. Operating conditions were varied, with superficial gas velocity of up to 7.2 ms^{-1} at 60 °C riser temperature, and riser net solids mass flux up to 53·kgs^{-1}·m^2. The overall solids concentration was around 2-2.5% and about 10% in the lower part of the riser. Ozone concentration profiles, axial pressure profiles, solids net mass flux as well as local particle velocities were measured. Gas velocity profiles were assumed in order to calculate the total transport of ozone at different heights in the riser. The importance of the lower part of the riser with respect to mass transfer is shown.

INTRODUCTION

Circulating fluidized beds (CFB) have been used successfully for several years in processes where gas-solids mass transfer is of significant importance [1]. Examples are combustion of low-grade fuels, mineral processing, fluid catalytic cracking and as a catalytic reactor for the production of a number of speciality chemicals. Hence, a comprehensive knowledge of the mechanisms of gas-solids mass transfer is an important design factor. Although many works have been done in modelling gas-solids mass transfer, *the understanding of the process of the mass transfer in a circulating fluidized bed is still in a developing stage.* CFBs are typically characterised by significant internal mass transfer limitations causing strong radial concentration gradients of gaseous reactants. This may be attributed to the existence of solids clusters and strands within the riser. In a CFB there is a high solids volume fraction in the bottom zone and a low one in the upper zone. Gas-solids fluid dynamic conditions can vary widely so that local solids concentration may cover the whole possible range; from minimum fluidization concentration φ_{mf} to concentrations approaching single particle flow. Up to now no gas-solids mass transfer model exists that is applicable for both low and high solids volume fractions. CFB gas-solids mass transfer cannot be related satisfyingly to gas-solids relative velocities occurring in CFBs due to unknown radial distribution of gas-solids relative velocity, axial solids back-mixing, radial mixing of solids between centre and wall regions.

To overcome this gap a series of experiments on gas-solids mass transfer coefficients, solids concentrations and solids velocities was performed to obtain a basis for a more reliable understanding of the occurring mechanisms. A number of previous studies are known concerning mass transfer and fluid dynamics within fluidized bed systems, where the decomposition of ozone has been used as a model reaction. Also for CFB systems mass transfer investigations have been carried out where decomposition of ozone has been applied. Published studies that are known to the authors are summarised in Table 1.

MASS TRANSFER IN A CFB

The decomposition of ozone is a first-order reaction. Hence, the reaction rate constant is independent of the gas concentration. The reaction takes place at low temperatures (ambient), and is therefore well suited for mass transfer studies in experimental set-ups. The external mass transfer in terms of global rate is to equate the mass transfer from bulk gas to catalyst surface:

$$\beta_v a_v (C_B - C_s) = \eta K_{int} \rho_s C_s \tag{1}$$

Eliminating C_s from Equation (1) and expressing $\mathbf{r}_p$ in terms of C_B gives

$$\mathbf{r}_p \rho_s = \frac{1}{\frac{1}{\beta_v a_v} + \frac{1}{\eta K_{int} \rho_s}} C_B = \beta_0 a_v C_B , \tag{2}$$

[1] Norwegian University of Science and Technology, Norway
[2] F. Hoffmann-La Roche AG, Switzerland

Table 1 Studies of CFB mass transfer using ozone decomposition

		[2]	[3]	[4]	[5]	[6]	[7]	this work
Riser material		PVC	N.A.	Stain. steel	Plexiglas	Stain. steel	N.A.	St-37
Riser diameter	mm	100	50	254	102	254	400	411
Riser height	m	5	4.5	10.85	6.32	10.85	15.6	8.5
Superficial velocity	$m \cdot s^{-1}$	3	3.8-8.8	2.27	1.5-2.5	3.8; 3.9	2.5-4.6	5.6-7.2
Temperature	°C		15	N.A.	23	N.A.	N.A.	60
Solids net flow rate	$kg \cdot m^{-2} \cdot s^{-1}$	15; 80	20.4-102	54.1	5.8-28.9	34; 106	11-42	31-53
Type of catalyst		FCC	γ-alumina with ferric oxide	FCC with ferric nitrate	FCC with ferric oxide	FCC with ferric nitrate	K-306; aluminium hydrosilicate	Angular cast steel
Particle size ($d_{50,3}$)	μm	46	82 (Sauter)	65	65	65	50	134
Particle bulk density	$kg \cdot m^{-3}$	2300	2970	1380	1500	1380	1420	3320
Reaction rate (k')	s^{-1}	N.A.	N.A.	3.956	4.72-5.52	14.8-57.2	1.5-6	26-62

where β_0 is the global apparent mass transfer coefficient based on solids volume. For small particles (for example $d_{50,3} \approx 100$ μm), the typical value of the Thiele modulus is low so that η approaches unity. This means that the total mass transfer needs only to be related to the bulk concentration of ozone in the air and not to the surface concentration.

The overall ratio between inertia forces and viscous forces in a CFB is larger (high Reynolds number) than in fixed beds, fluidized beds and in transport reactors. This means in principle high heat and mass transfer coefficients. Because a CFB is characterised by significant axial and radial gradients for gas velocity, solids velocity, solids concentration, one should distinguish between two types of mass transfer limitations.

1) In a system where the gas-to-particle mass transfer coefficient is much higher than the particle's ability to react the transported component,

$$\frac{1}{\beta_v a_v} << \frac{1}{\eta K_{int} \rho_s} \quad (3)$$

there are radial concentration profiles in the riser of the transported gas component (high solids concentration $\varphi \Rightarrow$ low concentration C, and low $\varphi \Rightarrow$ high C). The limitation for the gas-to-particle mass transfer is then radial gas mixing in the riser (*global transport*), and not that of the boundary layer around single particles (*local transport*). The meaning of the terms *global* and *local* is illustrated in Figure 1. For most CFB combustion processes the gas-to-particle mass transfer limitation actually is the radial gas mixing. One can also say that the limitation is caused by poor radial distribution of solids.

2) On the other hand, if the gas-to-particle mass transfer coefficient is much lower than the particle's ability to react the transported component,

$$\frac{1}{\beta_v a_v} >> \frac{1}{\eta K_{int} \rho_s} \quad (4)$$

the radial gas mixing in the riser is sufficient to avoid sharp radial concentration gradients. The value of the solids concentration φ is not so important (not limiting) because there is sufficient mass transfer area even at low solids concentration φ.

For mass transfer measurements in a CFB (for example ozone decomposition) one should be aware that when Equation (3) is fulfilled, one is actually measuring the gas mixing capability of the CFB system. This measured mass transfer coefficient is then something quite different from the perception of a mass transfer limitation related to a boundary layer. In a CFB one may have both conditions in 1), Equation (3), and 2), Equation (4), fulfilled at different positions in the CFB. From this statement a problem with mass transfer measurements arise; the mass transfer coefficient becomes dependent on the activity of the solids reacting the transported gas component. In principle, one should then make measurements both with an inert gas component (pure gas mixing) and with a reactive gas component. The difference between such two measurements would then give information about the transport related to the particle boundary layer. In order to describe the mass transfer of ozone on the CFB riser, an activity k (or K) which includes the various steps of mass transfer, is used in an expression based on a PFR behaviour of the CFB riser.

$$u\frac{\partial C}{\partial z} + \mathbf{r}_p \rho_B = u\frac{\partial C}{\partial z} + k\,C = u\frac{\partial C}{\partial z} + K\,\rho_s\,\varphi\,C = 0 \quad (5)$$

EXPERIMENTAL SET-UP

The CFB experimental rig used in the present work is (see Figure 2) is a semi-industrial plant being operated at low temperatures (from ambient temperature up to 70-80 °C). It consists of a 8.5m high circular riser with an inner diameter of 0.411m, a cyclone with a high efficiency (> 99%), a circular downer with inner diameter 0.324m, and at the bottom of the plant an quadratic siphon connecting the downer and the riser. Two Roots blowers with a maximum capacity of 6600 m^3/h supply primary and

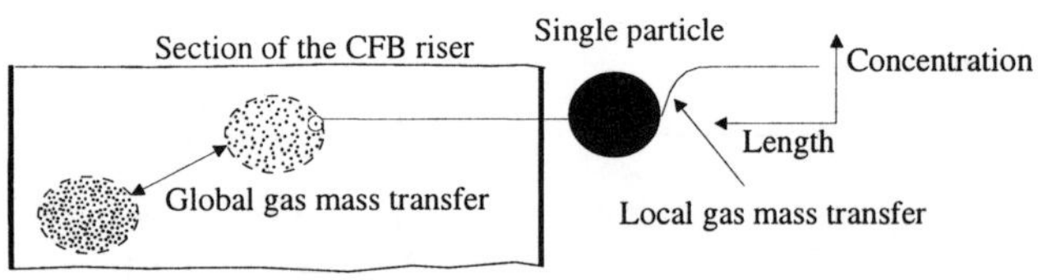

Figure 1 Illustration of the difference between the terms "global" and "local" gas mass transfer

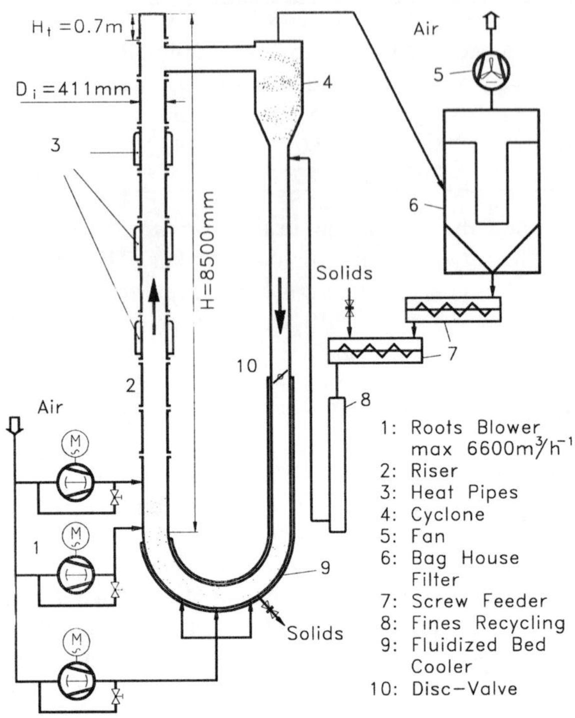

Figure 2 The Circulating Fluidized Bed at ETH Zürich

secondary air for the riser. Additionally, another blower as well as a compressed air facility supply air for fluidization at some points at the downer and in the siphon. The solids leaving the cyclone together with the air is collected in a bag filter, and returned periodically to the downer. This is, however, only a very small flow of solids compared to the flow in the CFB. In the downer there is a so-called disc-valve that can be closed in order to measure the solids net flow rate in the CFB. For a more detailed description of the plant, see [8, 9, and 10].

In the present work, ozone was generated with a device based on the corona discharge method was used (Ozonia CFS-1). Angular cast steel particles (SGT - ESGG200) were used as solids in the CFB. This material is typically used in grit blasting. The elemental analysis is: ≈96% Fe, 1% C, 1% Si, 1% Mn, 0.03% P and 0.03% S. Relevant properties for this study are given in Table 2. An X-ray spectrum analysis, using a scanning electron microscope, was carried out and it was established that there was oxygen in the layer which means there is an oxide. This is probably some form of iron oxide, which decomposes ozone.

An ozone-analyser from Monitor Labs, Inc. (Model 8810), using absorption of UV radiation as principle, was used to measure ozone concentrations. Gas samples from the CFB riser were continuously collected and analysed by the experimental set-up shown in Figure 3. The gas-sampling probe consists of a stainless steel tube with a PTFE (Teflon) tube inside. On the tip of the probe a disc of sintered bronze (SGT AG, Germany) is put inside the PTFE tube. The disc (d=11mm) was pressed inside the smaller PTFE tube (d_i =10mm) after blocking it out at elevated temperature. The PTFE tube with the Disc was then pulled back into the steel tube so that around the disc and between the PTFE and steel tube it was no leakage of solids. The sintered bronze disc has a retention diameter of 15μm, and it showed to have a good retention effect for the solids. After some hours of operation some fines could be observed inside the PTFE tube.

The discs were clogged with fines after a period of operation with a corresponding pressure drop which was measured by the ozone analyser. From time to time the probe was flushed with pressurised dry air through the tee, and this had a significant cleaning effect on the sintered bronze disc with respect to pressure drop. It was not observed that this cleaning influenced the measured concentration of ozone. This may be explained by the fact that the sample gas residence time through the disc is very short. The sintered bronze disc and the PFTE tubes were also cleaned periodically with methanol. The sampling probe was led into the riser through plastic ball valves. There are very small clearances between the sampling tube and the valve, which proved to offer an efficient sealing. Small marks on the outside of the steel tube were made in order to adjust the CFB riser radial position of the tip of the probe. At the sample gas inlet port of the ozone analyser, a 5 μm paper filter was applied. An air pump was used to suck a constant flow of sample gas through the probe. The pump was located downstream of the ozone analyser in order not to contribute with decomposition of ozone. A flow meter at the ozone

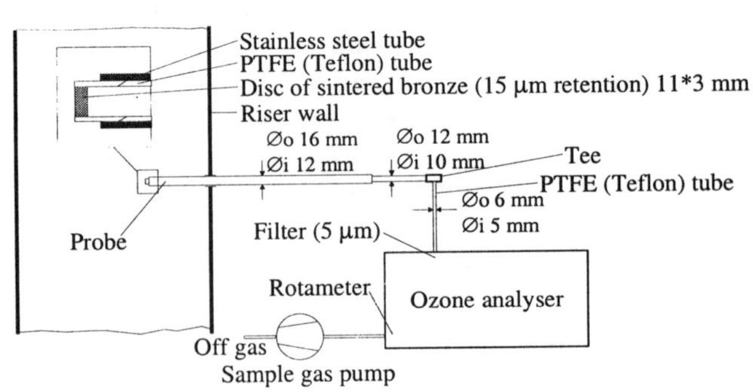

Figure 3 Experimental set-up for ozone concentration measurements.

Table 2 Properties of the *ESGG200* with particle size distribution curves

Property		Value
Particle density	$kg{\cdot}m^{-3}$	7300
Bulk density	$kg{\cdot}m^{-3}$	3320
Average particle diameter[1], $d_{50,3}$	µm	134
Average particle diameter, $d_{10,3}$	µm	77
Average particle diameter, $d_{90,3}$	µm	204
Sauter particle diameter, d_S	µm	119
Archimedes number Ar^2, $d_{50,3}$		455
Archimedes number Ar, $d_{10,3}$		86
Archimedes number Ar, $d_{90,3}$		1604
Terminal falling velocity v_{tf}, $d_{50,3}$	$m{\cdot}s^{-1}$	1.8
Terminal falling velocity v_{tf}, $d_{10,3}$	$m{\cdot}s^{-1}$	0.83
Terminal falling velocity v_{tf}, $d_{90,3}$	$m{\cdot}s^{-1}$	3.0
Particle/air density ratio (60 °C, 1 bar)		7000
Hardness Rc		48-52

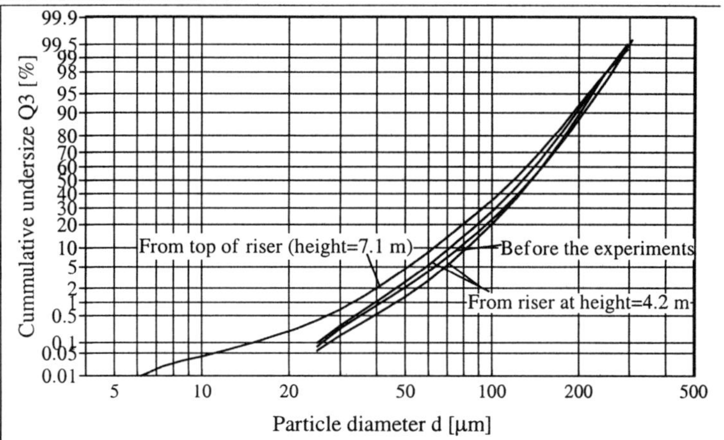

[1] The values for $d_{xx,3}$ are averages from 4 measurements using a laser analyser, Sympatec GmbH, Germany, [2]Calculated for 1 bar air pressure and 60 °C

Table 3 Measuring heights and volumes in the riser of the large CFB

Measuring point	Height z [mm]	Section	Height Δz [mm]	Volume [m^3]
H0	150	H0→H1	1390	0.1844
H1	1540	H2→H1	2670	0.3542
H2	4210	H3→H2	2870	0.3808
H3	7080	H3→H0	6930	0.9194

analyser exit port was used to control the flow of sample gas. The sampling tube flow rate was 0.5 l/min.

Solids velocity was measured with an optical probe using *light back scattering* [11, 8]. The light of two laser diodes is focused into a pair of optical fibres leading to probe tips. The lenses are bundling the light to form a defined measurement volume. Solids passing this probe volume reflect a certain amount of light. A second pair of optical fibres collects the reflected light to photo diodes. The light signals are converted into electrical signals which are further amplified to a range of 0-10 V. The analogue, high frequent electrical signals are stored in digital format on a personal computer for subsequent data processing. Ozone concentrations were measured at three heights in the riser (H1-3). The flow of ozone fed into the CFB was measured. The flow rate of ozone from the ozone generator could also be calculated from data provided by the manufacturer which were based on measured air flow and power input. The ozone generator showed some variation in the ozone production from day to day. According to information from the manufacturer the method of calculating the flow of ozone was not quite precise. Therefore the concentration of ozone was measured in the primary air admission tube. The height H0 is at the primary air inlet port. In Table 3 geometrical information is given for the measuring heights. At each height of the riser (H1-3) ozone concentrations were measured at seven radial positions. Measurements were performed with five different operating conditions as given in Table 4.

In order to find the total flow of ozone in the riser based on concentration measurements in a cross-section, it is necessary to know the total flow of gas as well as the velocity profile. Based on measurements by [8, 12], local gas velocity profiles were assumed using the following assumptions: 1) The total flow of air going into the CFB riser was measured, and must be equal to $\int u\, dA_c$, 2) At the wall the velocity has a given negative value for the whole circumference, 3) The gas velocity profile is symmetric around the centre-line of the riser, 4) The $u(r)$ is monotonous. At a given cross-section of the riser the total flow of ozone can be calculated as

$$\dot{V}_{O_3} = \int_A y_{O_3} u\, dA_c \quad . \tag{6}$$

When working with ozone decomposition, the riser gas temperature and humidity is of importance with respect to the adsorption of ozone at the surface of the particles used. This was examined by [13], who found from packed-bed step-response tests with ozone that is was possible to reduce the adsorption of ozone on silica by increasing the temperature and relative humidity. In the present work a riser temperature of 60 °C was used, and the air relative humidity was varying typically in the range 7-9% at this temperature. The low humidity indicates that adsorption would take place, while on the other hand, the high temperature would reduce the adsorption. However, the temperature was kept constant at 60 °C and the variation in air humidity was so small that changes in air conditions did not influence the rate of ozone decomposition.

Table 4 Operating conditions for the experiments (solids inventory 950 kg)

Experiment no.	T_{riser}	Superficial air velocity	Primary Air flow (p)	Secondary air flow (s)	p/s[1]	Solids flux[2]	Solids mean residence time
	[°C]	[m·s^{-1}]	[m^3·h^{-1}]	[m^3·h^{-1}]	[-]	[kg·s^{-1}·m^{-2}]	s
3-97	60	5.6	416	2188	0.190	32	36
4-97	60	6.5	412	2665	0.155	31	36
8-97	60	5.6	418	2216	0.189	40	30
9-97	60	6.7	456	2703	0.169	53	23
11-97	60	7.2	463	2890	0.160	46	28

[1] The ratio between primary and secondary air flow rates. [2] Referred to the riser cross-sectional area.

RESULTS AND DISCUSSION

The total decomposition in blank runs was measured to approximately 2%, of which most took place in the primary air inlet ports. The ozone decomposition in the sampling tube and in the sintered bronze disc at the tip of the sampling probe was negligible. Mixing of ozone with air in the primary air inlet was found very good. The typical maximum concentration of ozone was approximately 5 ppmv.

The solids showed a transient behaviour with respect to decomposing ozone after starting up the CFB. Very small concentrations of ozone in the riser were found at the start-up of operation. For a few hours the concentration at a certain point in the riser increased gradually to a stable level. From this transient behaviour, it can be concluded that locally in the riser (considering single particles), the limiting step of ozone mass transfer is the activity and not the convective/diffusive transport onto the surface of the particle. That means the condition in Equation (3) is fulfilled, meaning that the radial gas mixing is actually described by the measurements. This statement is based on the fact that mass transfer decreases in the transient period when operating condition of the CFB (air velocity, temperature, solids net flux, etc.) is kept constant. The conditions for mass transfer external to the particles will then be constant, and the only thing that may change is the phenomena at the particle surface.

There may be three different reasons for this transient behaviour of the solids' ability to decompose ozone; a) changes of the particle surface structure, b) changes of the particle surface composition, and c) a changing rate of adsorption of ozone on the particle surface due to surface concentration of ozone. The results given in the following are obtained in steady-state operation of the CFB, including that axial concentration profiles of ozone are developed. In this situation the "consumption" of ozone is looked at as a lump without distinguishing between the different steps involved. Important here is that the activity of the solids remains constant over time, and that the ability of the solids to "consume" ozone is expressed as a lump parameter, namely the activity. The mechanisms, as given above as a)-c), may change, but this does not matter as long as the sum of mass transfer resistance or ability to "consume" ozone remains constant.

In Figure 4 radial profiles for ozone concentration are shown. The profiles have a common form; meaning that there is very little ozone being measured at the wall, while in the centre of the riser the highest concentration was measured.

Measurements of solids velocities were carried out at the same three riser heights as for the ozone concentrations measurements. Radial profiles were measured (see Figure 5). The solids mass flux was measured with a disc-valve in the downer and is given in Figure 5. The solids flux is related to riser cross-sectional area.

Characteristic for the solids velocities shown in Figure 5 is the negative velocity (downward falling particles) of the solids near the wall. This could also be observed through inspection glasses in the riser wall. In the centre of the riser solids velocities around 4-6 m·s^{-1} was measured, which is somewhat below the riser superficial gas velocity ranging from 5.6-7.2 m·s^{-1}. In the centre of the riser at height 4.2 and 7.1m, the solids concentration is very low and the evaluation procedure of the data from the optical probe with these conditions was found not reliable. When interpolating the velocity in the centre of the riser in Figure 5, one obtains a maximum solids velocity approximately 1.5-2 times the superficial gas velocity. In the same experimental rig, [8] has measured this factor close to 2 with glass beads, using the same optical probe. The pressure profiles in the riser was measured, and it was found that for the different experimental operating conditions that most of the solids is at the lower end of the riser (see Table 5). Such profiles are typical in CFB combustion. The density ratio between solids and air was 7000 (1 bar and 60 °C).

There seemed to be no clear dependency between the superficial gas velocity and the solids net mass flux. During the experiments it was observed that the amount of fluidization air at the bottom of the CFB had a significant impact on the solids flow. The reason for this was probably that the inventory in the CFB was rather low. In Figure 6 the reaction rate constants, per unit mass of solids and per unit volume of riser, vs. superficial gas velocity, net mass flux of solids and solids concentration are depicted. The definitions of the reactions rate constants are given in Equation (5).

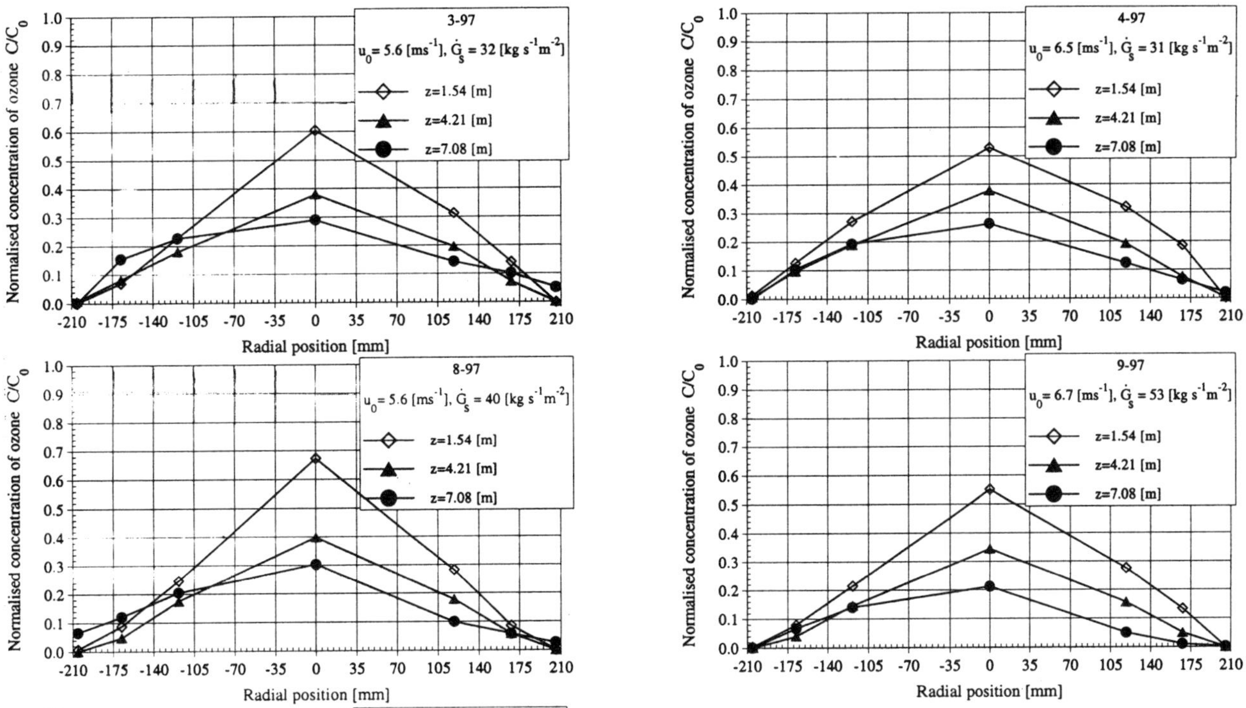

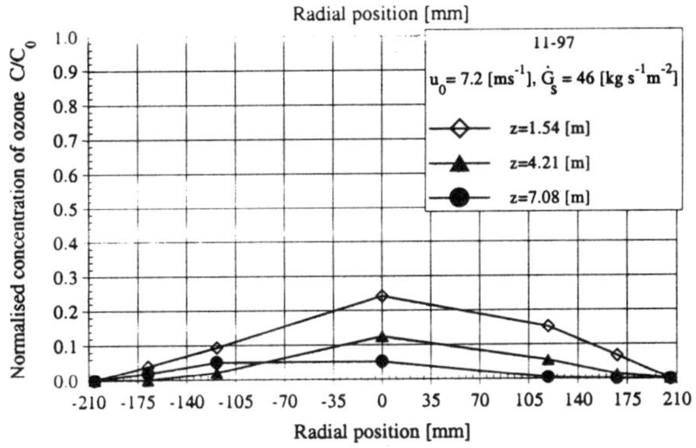

Figure 4 Radial profiles of ozone concentration in the CFB riser at three different riser heights.

Table 5 Mean solids concentration and mass of solids between the measuring points

	Units	3-97	4-97	8-97	9-97	11-97
Solids concentration φ, H0→H1	m^3_{solids}/m^3_{riser}	9.9%	9.8%	10.4%	10.2%	10.3%
Solids concentration φ, H1→H2	m^3_{solids}/m^3_{riser}	0.35%	0.36%	0.43%	0.63%	0.74%
Solids concentration φ, H2→H3	m^3_{solids}/m^3_{riser}	0.22%	0.25%	0.27%	0.39%	0.45%
Solids concentration φ, H0→H3	m^3_{solids}/m^3_{riser}	2.2%	2.2%	2.4%	2.5%	2.5%
Mass of solids, H0→H1	kg	134	132	140	138	138
Mass of solids, H1→H2	kg	9.1	9.2	11.2	16.4	19.1
Mass of solids, H2→H3	kg	6.1	7.0	7.4	10.8	12.6
Mass of solids, H0→H3	kg	149	148	159	165	169

There seems to be a correlation between superficial gas velocity and the reaction rate constants. For the reaction rate per unit volume of riser, the lower zone in the riser has the highest reaction rate, while the upper zone has the smallest. This points out the mass transfer limitation, which is gas mixing, in the dilute phase of the CFB riser flow. In the lower part of the riser the gas-solids contact efficiency and mass transfer is clearly better. Per unit mass of solids, the lower zone in the riser (H1-H0) has the smallest reaction rate. This can be explained by the large concentration of solids where some of it is not active in the sense of not being exposed to the flowing gas. The lower part of the riser is more like a turbulent fluidized bed. One can question if the incoming ozone gas is well mixed radically in the lower part of the riser. Measurements without any solids showed very good radial mixing of ozone.

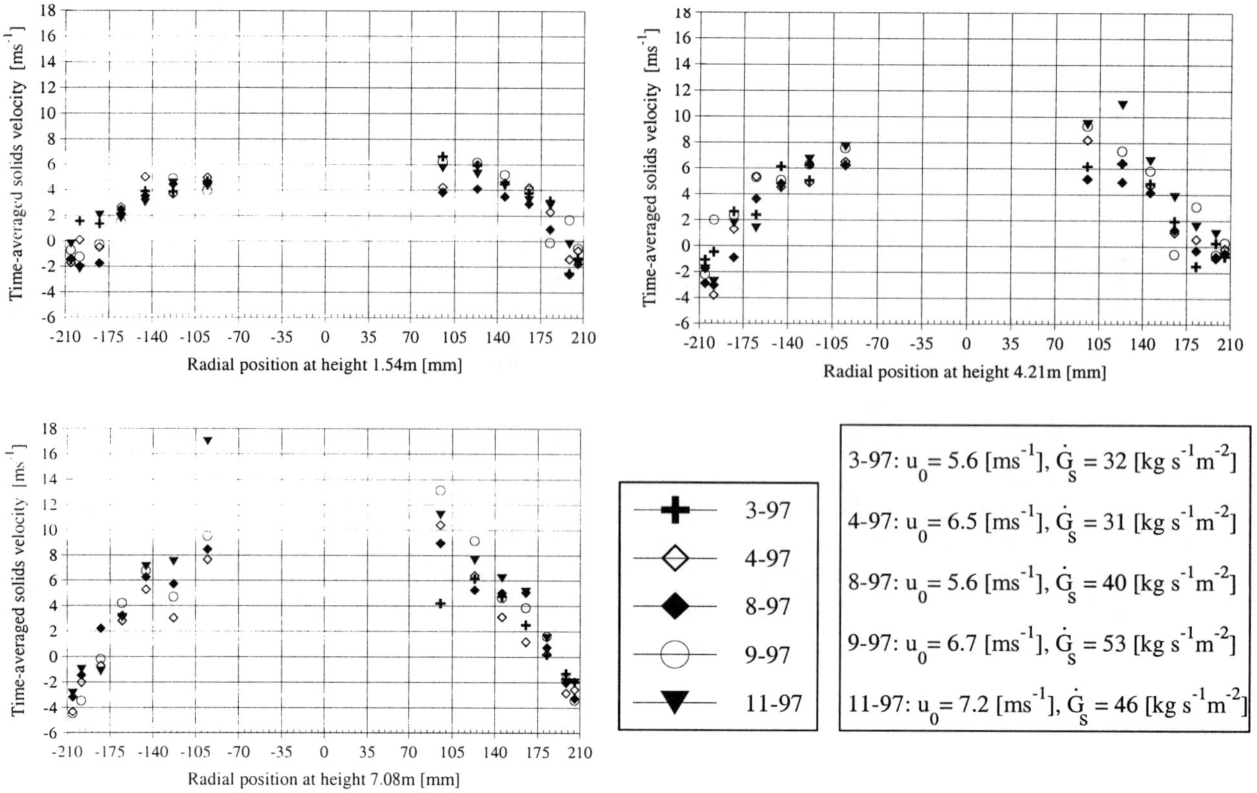

Figure 5 Time-averaged solids velocity profile in the CFB riser at three heights. Superficial gas velocities and solids mass fluxes are given.

The middle and the upper zone of the riser have a higher reaction rate per unit of mass solids. The results show that most of the ozone decomposition reactions take place in a rather small volume of the riser (lower part). However, the gas-solids contacting in this section of the riser is to some extent insufficient. This may be caused by non-uniform gas by-passing through the lower part of the riser. It is important to carefully design the inlet ports for both primary and secondary air (or feeding gas) in order to improve the gas-solids contacting (see also [14]). There seems to be no or very little correlation between the net solids mass flux and the reaction rate. One can clearly see the much stronger dependency between reaction rate and gas velocity than for net solids mass flux. This is somewhat contradictory to other studies (see references in Table 1). One major difference between those studies and the present work is that the superficial gas velocity and net solids mass flux are almost independent to each other in this study. There is more or less a linear relationship between reaction rate and solids concentration. This illustrates that the activity of the solids are the limiting step for decomposition of ozone, because the total particle surface area is proportional to solids concentration for a given volume of the riser. It should be noted that these results are to some extent specific for the CFB plant where the experiments were carried out. However, it points out the importance of making experimental CFBs as similar as possible to industrial plants for the lower part of the riser. In particular, having separate injection ports for air (primary/secondary) would be of importance. In the present study the Sherwood numbers obtained are in the range 10^{-5} to 10^{-3}. This is much lower than values found in the literature for example for fixed bed reactors. It is therefor obvious that the mass transfer limitation is related to gas mixing in the riser. The highest Sherwood numbers were found in the lower section of the riser, and the smallest ones in the upper zone of the riser.

CONCLUSIONS

Ozone concentration profiles (at 3 heights in the riser, and riser inlet), axial pressure profiles, solids net mass flux as well as local particle velocities were measured. Gas velocity profiles were assumed in order to calculate the total transport of ozone at different heights in the riser. Measuring local solids concentration, with the same optical probe, which was used for local velocities, was not successful. The experimental conditions were, with respect to solids-air density ratio and for the solids distribution in the riser, much like a CFB combustor with an overall solids concentration around 2-2.5% and about

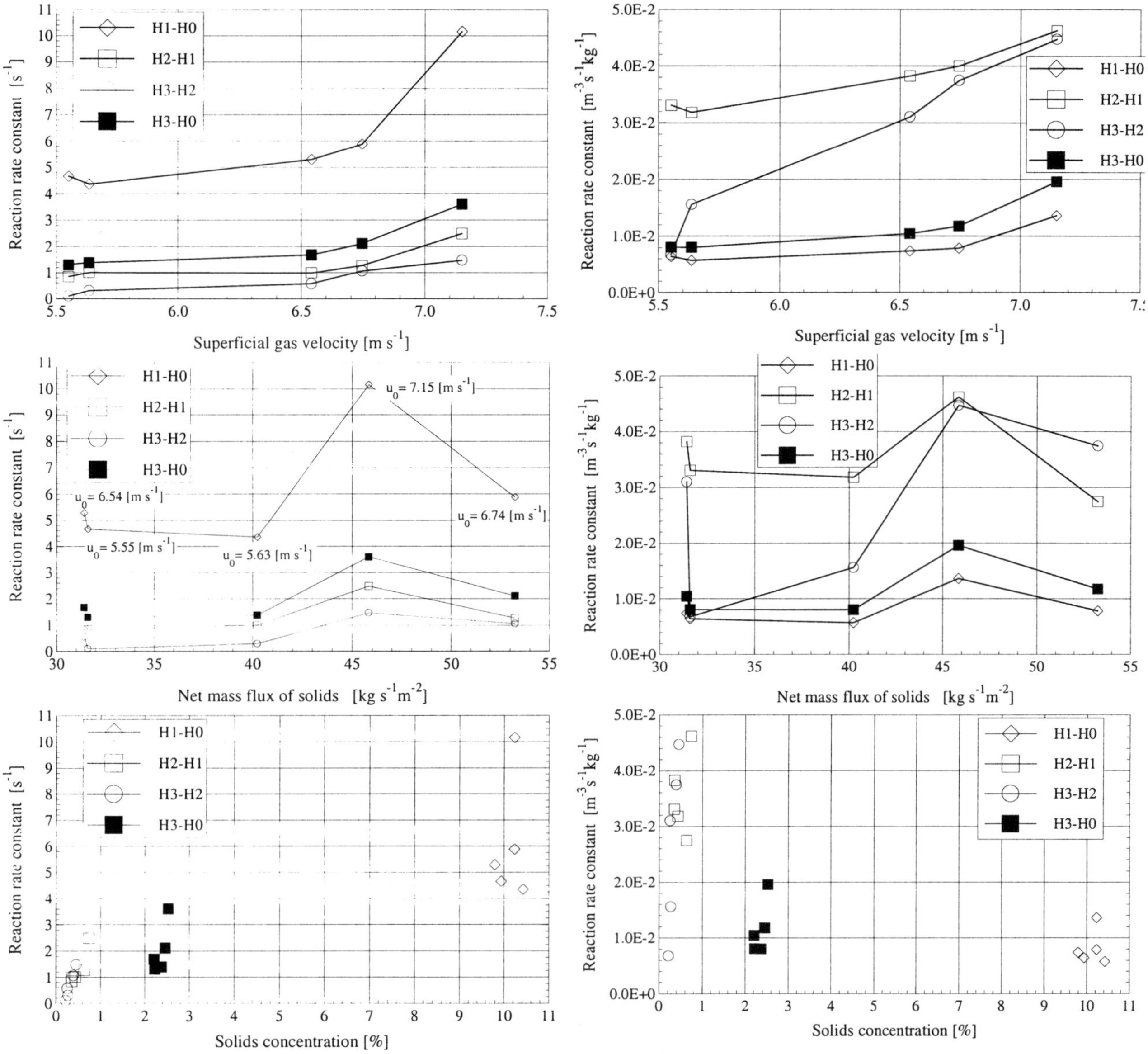

Figure 6 Reaction rate constants for different volume elements of the CFB riser. Upper left: (k) per unit volume of riser vs. superficial gas velocity (u_0). Upper right: (K) per unit mass of solids vs. superficial gas velocity (u_0). Mid left: (k) per unit volume of riser vs. net mass flux of solids ($\dot{G}_S$). Mid right: (K) per unit mass of solids vs. net mass flux of solids ($\dot{G}_S$). Lower left: (k) per unit volume of riser vs. solids concentration (φ). Lower right: (K) per unit mass of solids vs. solids concentration (φ).

10% in the lower part of the riser. The radial ozone concentration profiles have a common form; meaning that there is very little ozone being measured at the riser wall, while in the centre of the riser the highest concentration was measured. The profiles were almost symmetric around the centre-line of the riser. Characteristic for the local solids velocities is the negative velocity (downward falling particles) of the solids near the wall. This phenomenon could also be observed through inspection glasses in the riser wall. The results show that most [66-86% before height H1] of the ozone decomposition reactions take place in a rather small volume of the riser (lower part). However, the gas-solids contacting in this section of the riser is to some extent insufficient, because per unit mass of solids this section gave the lowest reaction rate. This can be explained by the large concentration of solids where some of it is not active in the sense of not being exposed to the flowing gas. It is therefore important, in order to obtain an efficient CFB reactor, to carefully design the inlet ports for both primary and secondary gas in order to improve the gas-solids contacting. For the reaction rate per unit volume of riser, the lower zone in the riser has the highest reaction rate, while the upper zone has the smallest. This is pointing out the mass transfer limitation, which is gas mixing, in the dilute phase of the CFB riser flow.

NOTATION

A_c	$[m^2]$	Cross-sectional area of riser, reactor
a_v	$[m^2 \cdot m^{-3} = m^{-1}]$	External surface per volume of catalyst
Ar	[-]	Archimedes number
C	$[mol \cdot m^{-3}]$	Concentration
C_0	$[mol \cdot m^{-3}]$	concentration of ozone when assuming mixing primary and secondary air without contacting the solids
C_B	$[mol \cdot m^{-3}]$	Concentration, bulk
C_s	$[mol \cdot m^{-3}]$	Concentration, surface
d	[m]	Diameter
$d_{50,3}$	[m]	Particle diameter (volume median)
d_S	[m]	Particle diameter (*Sauter*-mean value)
$\dot{G}_s$	$[kg \cdot m^{-2} \cdot s^{-1}]$	Solids mass flux
k	$[s^{-1}]$	Reaction rate constant, per unit volume of riser
k'	$[s^{-1}]$	Reaction rate constant, per unit volume of solids
K	$[m^3 \cdot kg_{cat}^{-1} \cdot s^{-1}]$	Reaction rate constant, per unit mass of catalyst
K_{int}	$[m^3 \cdot kg_{cat}^{-1} \cdot s^{-1}]$	Intrinsic reaction rate constant, per mass of catalyst
r	[m]	Radial position
$\mathbf{r}_p$	$[kmol \cdot kg_{cat}^{-1} \cdot s^{-1}]$	Reaction rate of component *i*, per kg catalyst
S	[-]	Dimensionless signal from the photo diode
T	[K]	Temperature
u	$[m \cdot s^{-1}]$	Gas velocity
u_0	$[m \cdot s^{-1}]$	Superficial gas velocity
v_{tf}	$[m \cdot s^{-1}]$	Particle terminal falling velocity
$\dot{V}$	$[m^3 \cdot s^{-1}]$	Volumetric flow rate
y_i	[-]	Mole fraction of gas component *i*
z	[m]	Axial position, riser height
β_0	$[m \cdot s^{-1}]$	Apparent mass transfer coefficient, based on solids volume
β_v	$[m \cdot s^{-1}]$	Mass transfer coefficient, based on solids volume
η	[-]	Effectiveness factor
φ	$[m^3_{solids} \cdot m^{-3}_{bed}]$	Solids concentration
ρ_B	$[kg \cdot m^{-3}]$	Density of bed of solids
ρ_s	$[kg \cdot m^{-3}]$	Density of solids, particle

LITERATURE CITED

1. **Reh, L.**, "New and Efficient High-Temperature Processes with Circulating Fluidized Bed Reactors", Chem. Eng. Technol., 18, pp 75-89, (1995)
2. **Kagawa H., Mineo H., Yamazaki R. and Yoshida K**, "A gas-solid contacting model for fast fluidized bed", in Circulating Fluidized Bed Technology, p. 551-556, (1990)
3. **Pagliolico, S., M. Tiprigan, G. Rovero and A. Gianetto**, "Pseudo-homgeneous Approach to CFB Reactor Design", Chemical Engineering Science, Vol. 47, No. 9-11, pp. 2269-2274, (1992)
4. **Ouyang, S., J. Lin and E. Potter**, "Ozone decomposition in a 0.254 m diameter circulating fluidized bed reactor", Powder Technology, 74, pp. 73-78, (1993)
5. **Jiang, P., H. Bi, R. Jean and L. Fan**, "Baffle Effects on Performance of Catalytic Circulating Fluidized Bed Reactor", AIChE Journal, Vol. 37, No. 9, (Sep. 1991)
6. **Ouyang, S., X.-G. Li and O.E. Potter**, "Circulating Fluidized Bed as a Catalytic Reactor: Experimental Study", AIChE Journal, Vol. 41, No. 6, (1995)
7. **Schoenfelder, H., M. Kruse and J. Werther**, "Two dimensional model for circulating fluidized bed reactors, AIChe-Journal Jul 1996; 42 (7) : pp. 1875-1888, (1996)
8. **Nicolai R.**, "Experimentelle Untersuchungen zur Strömungsmechanik in einer hochexpandierten zirkulierenden Gas/Feststoff-Wirbelschicht", Ph.D.-Thesis 11344, ETH Zürich, Switzerland, (1995)
9. **Shi, D.**, "Fluiddynamik und Wärmeübergang in einer zirkulierenden Wirbelschicht", Ph.D.-Thesis 11908, ETH Zürich, Switzerland, (1996)
10. **Hugi E.**, Ph.D.-Thesis ETH Zürich No. 12171, (1997)
11. **Reh, L. and J. Li**, "Measurement of Voidage in Fluidized Beds by Optical Probes", Circulating Fluidized Bed Technology III, Pergamon Press, Oxford, (1991)
12. **Karches, M.**, "Untersuchungen zur Strömungsmechanik in einer zirkulierenden Wirbelschicht", Diploma-Thesis, Universität Karlsruhe Germany/ETH Zürich Switzerland, (1997)
13. **van Swaaij, W.P.M. and F.J. Zuiderweg**, "Investigation of ozone decomposition in fluidized beds on the basis of a two-phase model", Proc. of the European Symp. on Chemical Reaction Engineering, Amsterdam, Elsevier, pp. B9-25, (1972)
14. **Steiner, C.P.**, "Thermische Verwertung von Abfallschlämmen in zirkulierender Wirbelschichtfeuerung unter Berücksichtigung des Einflusses der Aufgabe- und Bettkörnung", Ph.D.-Thesis No. 11888, ETH Zürich, Switzerland, (1996)

The ETH Experience: Experimental Database and Results from the Past Eight Years

Peter Herbert and Lothar Reh
Institute of Process Engineering (IVUK), Swiss Federal Institute of Technology (ETH), 8092 Zurich, Switzerland
Rainer Nicolai
Hoffmann-La Roche, CH-4070 Basel, Switzerland

Many papers have been published in recent years dealing with the modeling of the hydrodynamics of gas-solid systems in general and of circulating fluidized bed (CFB) risers in particular. The physics at work in these systems is very complex and has yet to be fully understood. Better co-operation is needed between experimental groups and theoretical research groups in order to pinpoint the strengths and weaknesses of current methodologies and to better define the direction in which future research should proceed.

At the Institute of Process Engineering at the Swiss Federal Institute of Technology (ETH) in Zurich detailed fluid dynamic, heat transfer, and reaction experiments have been performed on a semi-industrial scale CFB (40 cm I.D. and 8 m in height) over the past eight years. The results of this continuous and comprehensive campaign have been the subject of numerous publications and the data has been systematically stored in an Oracle database. In this paper the experience at ETH is reviewed and the contents and organization of the data base is discussed.

INTRODUCTION

Experimental work began over 8 years ago on a semi-industrial scale Circulating Fluidized Bed (CFB) at the Swiss Federal Institute of Technology in Zurich under the direction of Professor Lothar Reh. Since that time 6 doctoral theses have been published [1-6] along with numerous papers [e.g. 8-17] on topics varying from local measurements to heat-transfer and cyclone design. Local and innovative measurement techniques have been developed and have evolved with the changing priorities and objectives of the research projects. It was recognized from the outset that due to the constant flux of personnel in a university research laboratory that it was necessary to organize all of the data concerning the CFB operation in an easily accessible format. An experimental database was set up using Oracle and now contains data on all aspects of the CFB from the manufacturer of the pressure transducers to the measured radial profiles of particle velocity. By proceeding in such a systematic and organized manner unnecessary repetition could be avoided and new data could easily be analyzed and compared with previous measurements. Now that CFB research has terminated at ETH the experimental database will be made available to other interested groups.

EXPERIMENTAL EQUIPMENT

The current configuration of the CFB system at ETH Zürich is shown in Figure 1.

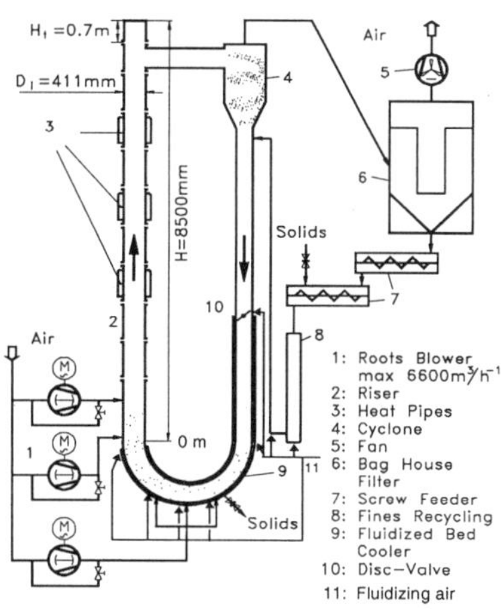

Figure 1: Cold model CFD unit at ETH Zürich - current configuration

The majority of the experiments that will be described however were performed using the arrangement in Figure 2.

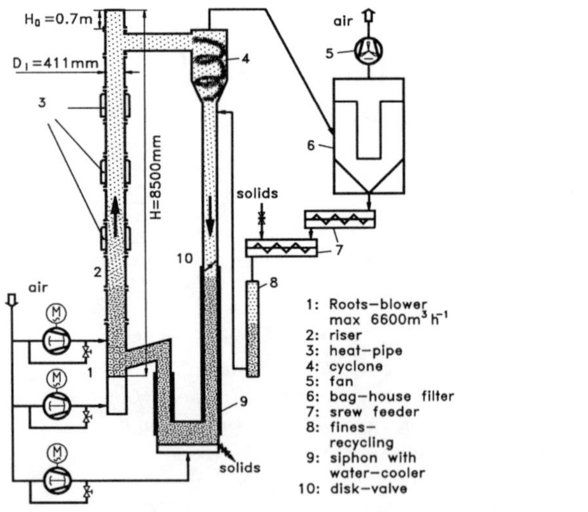

Figure 2: Cold model CFB-Unit at ETH-Zürich - prior configuration

This semi-industrial scale CFB has a riser inner diameter of 0.41 m and a height of 8.5 m. Air was supplied by three Roots-type blowers and the superficial velocity in the riser was set by adjusting the secondary air flowrate. The primary air flowrate was fixed at 570 m^3/h. The overall operating conditions were varied over the following ranges:

- *Superficial gas velocity: up to 13 m/s*
- *Solids circulation rate: up to 250 kg/m^{2}s*
- *Mean particle diameter (μm):* $42 < d_{p50,3} < 300$

Figure 3 shows the particle size distributions for the different fractions used.

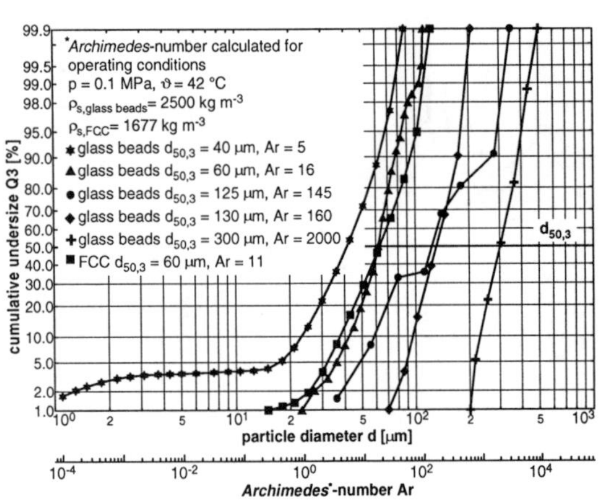

Figure 3: Particle size distributions

The solids circulation rate was determined directly by closing a disk valve in the downcomer and observing the change in the pressure drop across the descending bed of particles. An energy balance was also used to estimate the solids flux however this indirect technique was much less reliable than the disk valve method but could give a good indication of the stability of the system.

Approximately 100 measurement transducers automatically recorded all static pressures, temperatures, and flowrates at regular intervals (roughly every 3 minutes) during normal operation. This information was then transferred to the Oracle database. Local high frequency time series for particle concentration, velocity, and momentum determination are stored separately on CD ROM and the calculated average values of these parameters are available in the database.

Results and Discussion

It is impossible to describe all of the measurement techniques used and the results in detail here. Instead some representative figures will be presented for different operating conditions and the interested reader referred to other sources for more detailed information. A typical pressure profile for the CFB loop is shown in Figure 4.

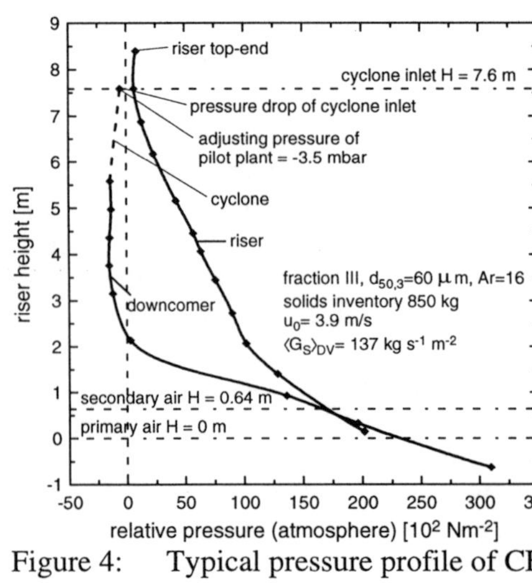

Figure 4: Typical pressure profile of CFB loop

The optical probe originally described by Reh and Li [7] has been constantly modified and improved over the course of the different doctoral works. With this system it is possible to measure local particle velocity and concentration simultaneously within a known measurement volume thus allowing reliable estimations of the local solids mass flux to be made. This probe has recently been adapted to allow measurements under high temperature conditions (850 °C) in the 250 MWel CFB of Electricité de France (EDF) in Gardanne, France [6].

Measurements in the 0.41 m diameter riser have shown a notable influence of the average particle diameter on the shape of the radial concentration and velocity profiles (Figs. 5, 6).

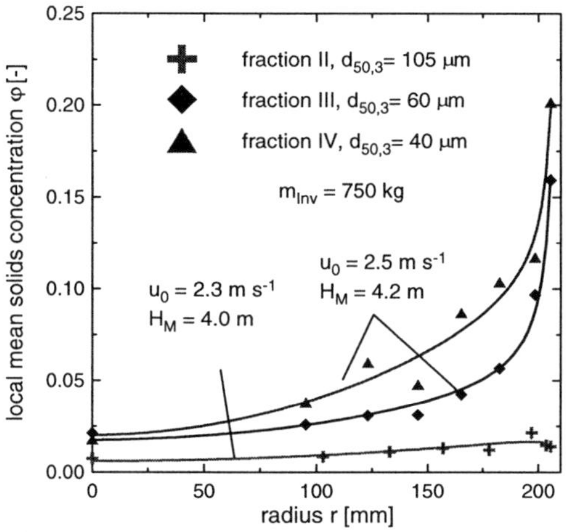

Figure 5: Radial particle concentration profiles for different particle diameters

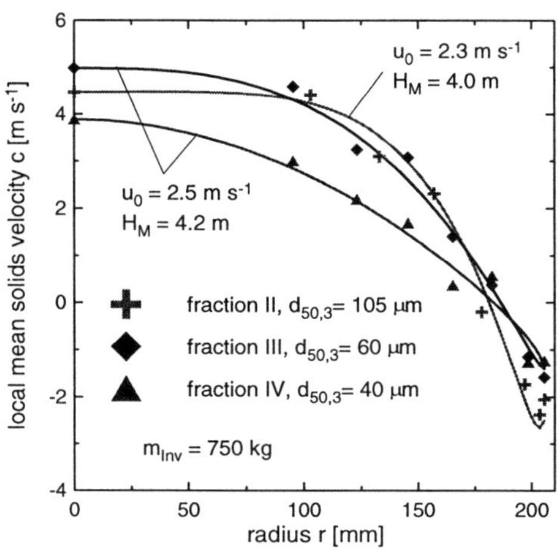

Figure 6: Radial particle velocity profiles for different particle diameters

The concentration profiles are much more uniform for the larger particle diameter whereas velocity profile shows an opposite tendency though the trend is not as significant statistically.

An independent method for determining the solids momentum flux was developed by Nicolai and described in [8]. Using a miniaturized piezo-resistive pressure transducer (1.4 mm x 1.4 mm) the pressure exerted by the momentum flux ($I_S = \rho_S \cdot c^2 \cdot \varphi$) could be accurately estimated after proper calibration [3]. The relative upward and downward fluxes measured simultaneously using this probe are shown in Figure 7.

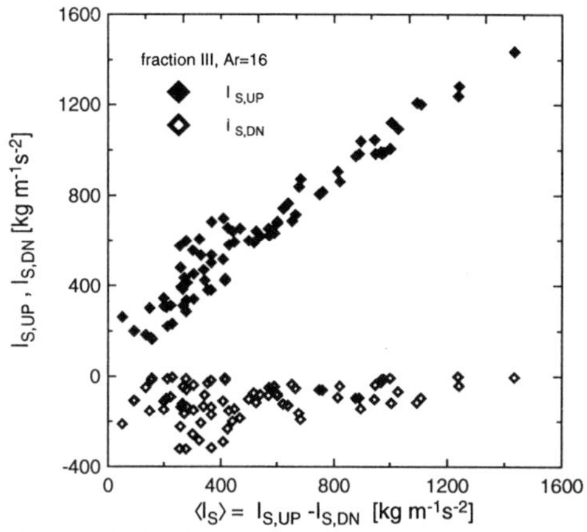

Figure 7: Relative upward and downward momentum flux in riser as a function of the overall flux

It is interesting to note that the downward momentum flux remains constant and only the upward momentum flux increases with increasing overall momentum flux.

Shi [4] performed extensive measurements of wall-to-bed heat transfer coefficients along with local particle velocity and concentration for the different particle size fractions shown in Figure 3. Three heat pipes with relatively large surface areas (0.74 m^2) for heat transfer were used to ensure that the measurements made would be representative of industrial situations. Figure 8 shows the effect of the average particle size on the heat transfer coefficient as a function of the particle concentration at the wall.

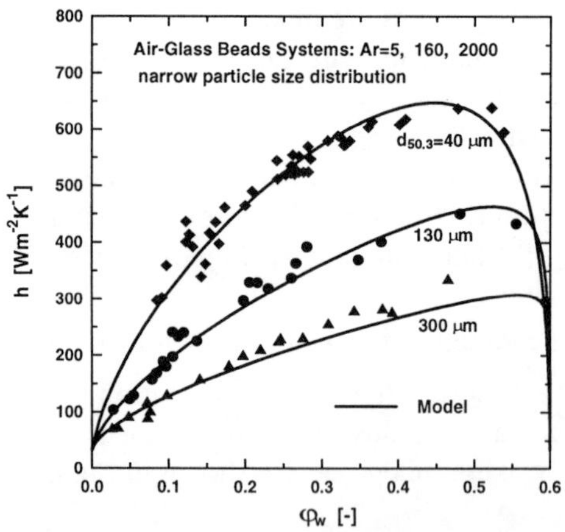

Figure 8: Wall-to-bed heat transfer coefficients as a function of particle concentration at the wall

The coefficients increase steadily with increasing particle concentration and with decreasing particle size. The model curve in this figure takes into account the time-averaged particle concentration at the wall as well as the magnitude of the concentration fluctuations through the standard deviation of the measured time series. These parameters were found to be the most important and the relative influence of the particle size distribution, superficial gas velocity, solids inventory and the height along the riser found to be minimal or nonexistent in comparison.

Summary and Conclusions

In this paper some important representative results have been presented from the past 8 years of CFB study at the Institute of Process Engineering at ETH Zürich. All of this information including operating conditions, stationary measurements, and technical details have been systematically stored in an Oracle relational database for ease of access and off-line analysis. The data represent several thousand operating hours of the semi-industrial scale CFB unit. A schematic diagram illustrating the contents of the database is given in Figure 9. It is our hope that this wealth of information will be used and evaluated by other groups so that the complex fluid dynamic behaviour of these systems can be better understood. A database of this type could be of considerable interest to those researchers involved in Computational Fluid Dynamic (CFD) modeling for evaluation purposes.

Work is currently being performed on the analysis of measured time series using techniques from chaos theory and on the application of neural networks for the optimization and control of CFB systems [18].

Acknowledgment

The authors gratefully acknowledge the financial support of the Electric Power Research Institute (EPRI), Palo Alto, the Swiss Federal Institute of Technology (ETH), Zürich, and the Swiss National Energy Research Fund, Basel.

Notation

$\bar{x}$ time average value

$\langle x \rangle$ cross sectional mean value includes time averaging

$$Ar = \frac{d_{50,3}^3 \cdot g}{\nu^2} \cdot \frac{\rho_s - \rho_g}{\rho_g}$$ *Archimedes*-number

c [m s^{-1}] solids velocity

$d_{50,3}$ [m] mass median particle diameter

d_p [m] particle diameter

g [m s^{-2}] gravitational constant

$\langle G_S \rangle_{DV}$ [kg s^{-1} m^{-2}] ext. solids mass flux metered by Disk-Valve

h [$Wm^{-2}K^{-1}$] heat transfer coefficient

H [m] height

H_M [m] height of measurement position

I_S [kg s^{-2} m^{-1}] solids momentum flux

$I_{S,UP}$ [kg s^{-2} m^{-1}] upward momentum flux

$I_{S,DN}$ [kg s^{-2} m^{-1}] downward momentum flux

m_{Inv} [kg] bed solids inventory incl. siphon

p [kg m^{-1} s^{-2}] pressure

r [m] radius

u_0 [m s^{-1}] superficial gas velocity

ε [-] gas volume concentration

$\varphi = (1-\varepsilon)$ [-] solids volume concentration

ρ_g [kg m^{-3}] gas density

ρ_S [kg m^{-3}] particle density

ϑ [°C] temperature

ν [m^2 s^{-1}] kinematic gas viscosity

LITERATURE CITED

[1] **Goedicke, F.,** "Strömungsmechanik und Wämeübergang in Zirkulierenden Wirbelschichten", ETHZ Diss. 9733, (1992).

[2] **Tanner, H.,** "Lokale Strömungsmechanik in hochexpandierten Zirkulierenden Gas/Festoff-Wirbelschichten", ETHZ Diss. 10805, (1994).

[3] **Nicolai, R.,** "Experimentelle Untersuchungen zur Strömungsmechanik in einer hochexpandierten Zirkulierenden Gas/Festoff-Wirbelschicht", ETHZ Diss. 11344, (1995).

[4] **Shi, D.,** "Fluiddynamik und Wämeübergang Strömungsmechanik in einer Zirkulierenden-Wirbelschicht", ETHZ Diss. (1996).

[5] **Hugi, E.,** "Auslegung hochbeladener Zyklonabscheider für zirkulierende Gas/Festoff-Wirbelschicht-Reaktorsysteme", ETHZ Diss. (1997).

[6] **Caloz, Y.,** "Experimental Investigation of Local Solids Fluid Dynamics in Different Industrial-Scale Circulating Fluidized Beds with Optical Probes", ETHZ Diss., *to be published in* (1998).

[7] **Reh, L., and J. Li,** "Measurement of Voidage in Fluidized Beds by Optical Probes", Proc. 3rd International Conference on Circulating Fluidized Beds, Nagoya, Japan, pp. 163-170, (1990).

[8] **Nicolai, R., and L. Reh,** "Measurements of Solids Concentration, Velocity and Momentum Distribution in a Cold CFB Unit", 13th Int. Conf. On Fluidized Bed Combustion, ASME, Orlando, Fl., (May 7-10, 1995).

[9] **Shi, D. and L. Reh,** "Wall-to-Bed Heat Transfer in Circulating Fluidized Beds", Proc. Of the 5th Int. Conf. On Circulating Fluidized Beds, Beijing, China, (May 28-31, 1996).

[10] **Nicolai, R. and L. Reh,** "Experimental Study on Cold Model CFB Fluid Dynamics", Proc. Of the 5th Int. Conf. On Circulating Fluidized Beds, Beijing, China, (May 28-31, 1996).

[11] **Reh, L.,** "Fluid Dynamics of CFB Combustors", Proc. Of the 5th Int. Conf. On Circulating Fluidized Beds, Beijing, China, (May 28-31, 1996).

[12] **Nicolai, R., and L. Reh,** "Experimental Investigation of Cold-Model CFB Fluid Dynamics", AIChE Annual Meeting, Miami, Fl., (Nov. 12-17, 1995).

[13] **Shi, D., Liu, K., Nicolai, R., and L. Reh,** "Fluctuations of Local Solids Concentrations and Velocities in a Circulating Fluidized Bed", Proc. Int. Symp. on Measuring Techniques for Multiphase Flows, Nanjing, China, (Apr. 10-13, 1995).

[14] **Nicolai, R., Zhoulin, Y., and L. Reh,** "Characterising Two-Phase Gas-Solid Flow in CFB by an Optical Probe", Proc. Int. Symp. on Measuring Techniques for Multiphase Flows, Nanjing, China, (Apr. 10-13, 1995).

[15] **Tanner, H., Li, J., and L. Reh,** "Radial Profiles of Slip Velocity between Gas and Solids in Circulating Fluidized Beds", AIChE Annual Meeting, St. Louis, (Nov. 7-12, 1993).

[16] **Nicolai, R., Goedicke, F., Tanner, H., and L. Reh,** "Particle Induced Heat Transfer Between Walls and Gas/Solid Fluidized Beds", Proc. of the 4th Int. Conf. On Circulating Fluidized Beds, Somerset, PA, (Aug. 1-5, 1993).

[17] **Goedicke, F. And L. Reh,** "Particle Induced Heat Transfer Between Walls and Gas/Solid Fluidized Beds", AIChE Annual Meeting, Miami, Fl., (Nov. 1-6, 1992).

[18] **Ye, H., Nicolai, R., and L. Reh,** "A Bayesian-Gaussian Neural Network and its Applications in Process Engineering - Modeling, Optimization and Control", Proc. of 23rd SPEEDUP Workshop on Process Engineering, Zurich, Switzerland, (March, 1998).

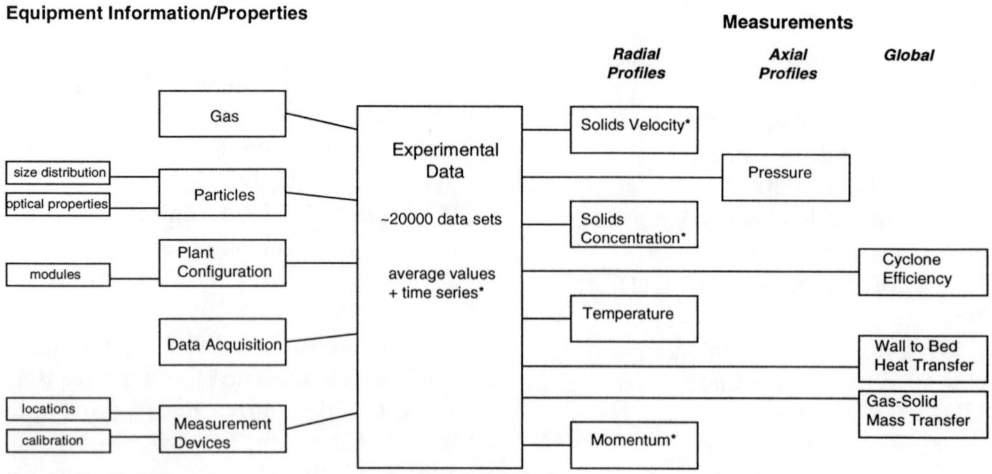

Figure 9: Schematic structure of relational database

Hydrodynamic Studies in a 0.45-m Riser with Liquid Feed Injection

David C. Skouby
Exxon Research and Engineering Company, 180 Park Avenue, Florham Park, NJ 07932

Hydrodynamic data were measured in a 0.45-m riser with FCC catalyst and liquid nitrogen feed. Catalyst concentration, catalyst flux, and particle velocity were measured using a variety of techniques including an extraction probe, a capacitance probe, and a fiber optic reflectance probe. The evolution of hydrodynamic profiles along the riser is tracked, and the effect of feed injector characteristics is examined.

Riser reactors are important in many industrial applications, particularly in fluid catalytic cracking (FCC) in the petroleum industry and in circulating fluid bed combustors (CFBC) in the power generation industry. Although the feed in FCC is generally liquid, the majority of experimental data are taken with gas feed into the riser. With liquid feed there is rapid evaporation in the feed injection zone that gives rise to an acceleration profile very different from that found with gas feed. This work examines the hydrodynamic behavior of risers with liquid feed injection.

EXPERIMENTAL

The experimental unit is a cold flow circulating fluidized bed, approximately 12 m. tall, which circulates up to 3 (T)/(min) of FCC catalyst. Data were taken in a 10.7-m. long, 0.45-m. ID riser. Feed was injected at the bottom of the riser using two types of feed injectors -- Nozzle 1, which has a conical spray pattern, and Nozzle 2, which has a flat spray pattern. Four nozzles are used at a time, spaced at 90° intervals around the circumference. A mixture of liquid and gaseous nitrogen is injected through the nozzles to contact hot catalyst circulating around from a 0.61-m. diameter dense fluidized bed downcomer, which is heated by steam coils.

Liquid nitrogen is used to simulate FCC feed, because it has several properties that make it behave similarly to the oil feed in an FCC unit. When liquid nitrogen is injected into the stream of hot catalyst in the riser, it evaporates and causes an increase in the amount of gas flowing up the riser, just as oil droplets evaporate and react in FCCU. In addition, the liquid nitrogen is partially flashed upstream of the injection nozzles in a controlled fashion to be able to simulate the two-phase injection of an oil/steam mixture. The liquid nitrogen also creates a temperature difference between the feed (at ~ 90 K) and the hot flowing catalyst (at ~ 310 K) similar to that found in FCCU's. Radial temperature profiles at different axial locations in the riser were measured, but the results are not reported here.

Catalyst Mass Flux Measurement Procedure

The mass flux probe is an extraction probe which operates on a simple "bucket and timer" principle. Catalyst is collected for a measured length of time and a catalyst mass flow rate is calculated. A mass flux is obtained by dividing the mass flow rate by the cross-sectional area of the probe mouth. To measure the catalyst flux at any location, the probe is inserted into the riser to the desired location, the probe valve is opened, the gas flow through the probe is adjusted with a rotameter, and the catalyst is collected for a desired time. Both upflow and downflow are measured at the same gas flow rate through the probe,

Exxon Research & Engineering, Co., Florham Park, NJ

measured at the same gas flow rate through the probe, and the net upward flux is calculated as the difference. Although the measurement is by necessity non-isokinetic (the gas velocity profile is *a priori* unknown), the net catalyst flux is reasonably approximated using this method [1-3]. Traverses were made across the riser, sampling at the far wall, 7/8-diameter, 3/4-diameter, centerline, 1/4-diameter, and 1/8-diameter. Sampling at the near wall was not done, because the curvature of the sampling probe would not allow the probe to be pulled flush with the near wall.

Particle Velocity Measurement Procedure

The particle velocity probe is a Vector Scientific Instruments® reflectance fiber optic probe. The probe consists of two bundles of fiber optics spaced a known distance apart and encased in a stainless steel sheath. The commercial software calculates average particle velocities by computing the cross-correlation between the two reflected signals, finding the lag time which gives the largest correlation, and dividing that into the distance between the fiber optic bundles. Data was taken at 12.5 or 25 kHz, 1048 samples were taken in each series, and five sets of 100 series were made for each probe location. Each series produces a correlated velocity, so a total of 500 velocity measurements were taken for each condition. As proposed by Kan, *et al.* [4] data points with a normalized cross-correlation coefficient less than 0.5 were rejected. Since the probe was only 0.3 m. long, traverses were made halfway across the riser, sampling at the centerline, at half-radius, at 3/4-radius, and at the wall.

Local Catalyst Concentration Measurement Procedure

The local catalyst concentration was measured using a capacitance/fiber optic probe designed and built by Professor Michel Louge at Cornell University. The probe consists of concentric capacitance and fiber optic probes to measure local catalyst concentration. The capacitance probe is calibrated on dry FCC catalyst using the method of Louge and Opie [5]. The calibrated capacitance probe and the fiber optic probe simultaneously measure the catalyst concentration at the wall of the vessel with and without solids flow. That information is used to calibrate the fiber optic probe, which is then traversed to get catalyst concentrations in the interior of the riser.

The operation of the fiber optic probe is described elsewhere [6]. In general the probe measures an optical reflectance signal which can be assumed to be linear with the catalyst concentration. Measurement of the optical signal in air and the optical signal at the wall of the riser (which corresponds to the catalyst concentration measured with the capacitance probe) defines the functional relationship between the optical voltage and catalyst concentration. The fiber optic probe is then traversed across the riser, yielding local catalyst concentration data across the riser diameter.

RESULTS

Hydrodynamic measurements were taken at 0.5 m., 0.91 m., and 4.9 m. above the injectors. Gas velocities after complete evaporation of the liquid nitrogen feed were approximately 9 m/s.

Mass Flux Probe Measurements

Results of the catalyst mass flux probe measurements are shown in Figure 1. The data are normalized to the integrated mass flux for each run.

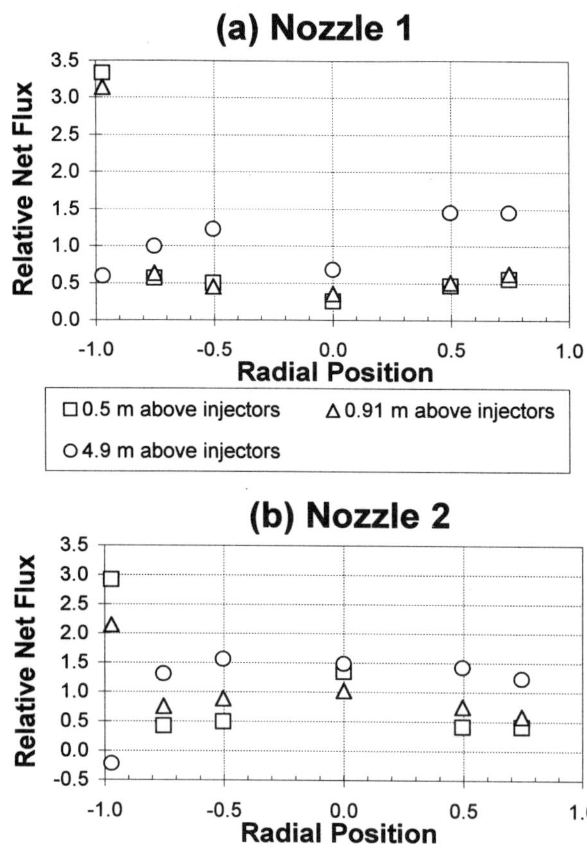

Figure 1: Radial net catalyst flux profiles for two different feed nozzles

The flux profiles appear to be fairly symmetric around the centerline, at least in the plane of the probe traverse. The integrated profiles agree with independently measured solids circulation rates to ±25%. The mass flux probe measurements revealed that the majority of the catalyst mass flux was at the wall close to the injection point and toward the center of the riser at the upper measurement. These results indicate that the liquid feed was forcing the majority of the catalyst away from the middle of the riser to the riser walls. For fully developed riser flow the maximum flux is expected to be at the middle of the riser, with negative flux at the wall (see *e.g.* [3]). Since the maximum flux is not at the centerline in Figure 1, 5 m. (11 L/D) appears to be insufficient distance under these conditions to reach truly fully developed flow in this riser. It is interesting to note that negative fluxes were seen at the wall for Nozzle 2, but not for Nozzle1. Given sufficient distance from the feed injection zone, the effect of the nozzles is expected to become negligible, so one would expect that the profiles should become similar. This is another indication that fully developed flow was not reached at the upper measurement position.

For Nozzle 1 (Figure 1a), the lower measurement positions indicated that the minimum catalyst flux was at the center of the riser. For Nozzle 2 (Figure 2b), the two lower measurement positions indicated a minimum of catalyst flux away from the centerline. These results indicate that the feed from Nozzle 1 penetrates to the center of the riser while the feed from Nozzle 2 penetrates only partway to the center, giving rise to dissimilar hydrodynamic profiles. Temperature variations in the riser are not expected to affect the results of this probe.

Particle Velocity Measurements

Results of the fiber optic particle velocity probe measurements are shown in Figure 2. These data were normalized to the centerline velocity at the upper measurement position for Nozzle 2. For Nozzle 1 (Figure 2a) the maximum particle velocity was always at the center of the riser. The centerline velocity initially increased and then decreased with distance from the injectors, indicating that the liquid feed may still have been vaporizing between the first two measurement positions. Particle velocity measurements for Nozzle 2 (Figure 2b) indicated that the maximum particle velocity at the two lower measurement positions occurred at a position off-center. These data confirm the notion that the two nozzles have different penetration characteristics. For Nozzle 2 the centerline velocity increased as the distance from the injectors increased. In all cases at the upper measurement position the maximum particle velocity is at the center of the riser. The particle velocity profiles become similar at the upper position, as would be expected. Temperature variations in the riser are not expected to affect these measurements, because the velocity is determined from a cross-correlation of signals instead of being dependent on the amplitude of the signals, which could vary with temperature.

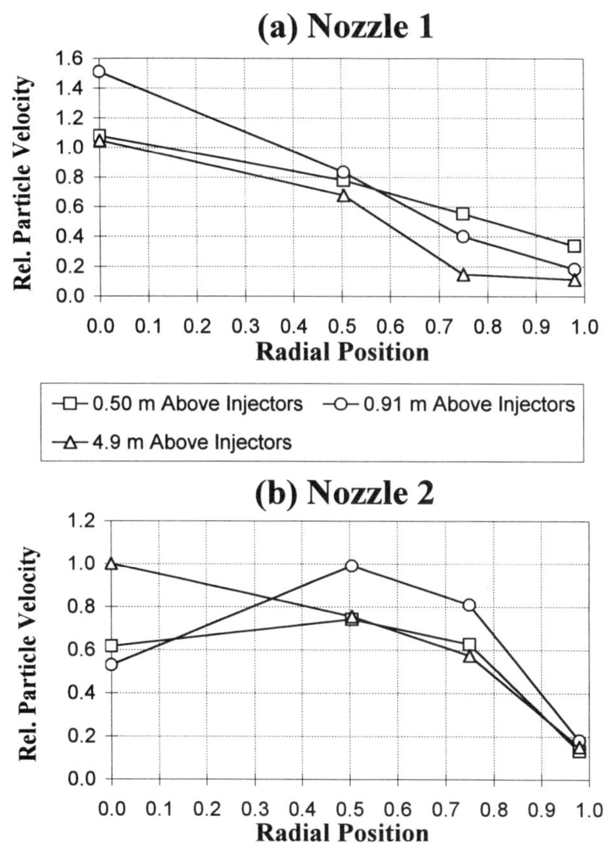

Figure 2: Radial particle velocity profiles for the two different feed nozzles

Local Catalyst Concentration Measurements

Plots of local catalyst fraction are shown in Figure 3. Values at the wall correspond to data from the capacitance/fiber optic probe. Values in the interior of the riser are calculated from the data for local catalyst flux and particle velocity. Measurements in the interior of the riser using the capacitance/fiber optic probe were suspect, because of drift in the optical voltage baseline, probably caused by extreme temperature variations within the riser. On the other hand, particle velocity measurements at the wall suffered from some apparent interference with the probe.

In all cases, the catalyst fraction at the wall is much greater than in the interior of the riser. For Nozzle 1 (Figure 3a) the catalyst fraction is lowest in the middle of the riser at all axial locations. For Nozzle 2 (Figures 3b) the catalyst fraction is lowest at approximately half-radius, except at the measurement point farthest from the injectors. Again, these results confirm the different penetration characteristics of the two nozzles.

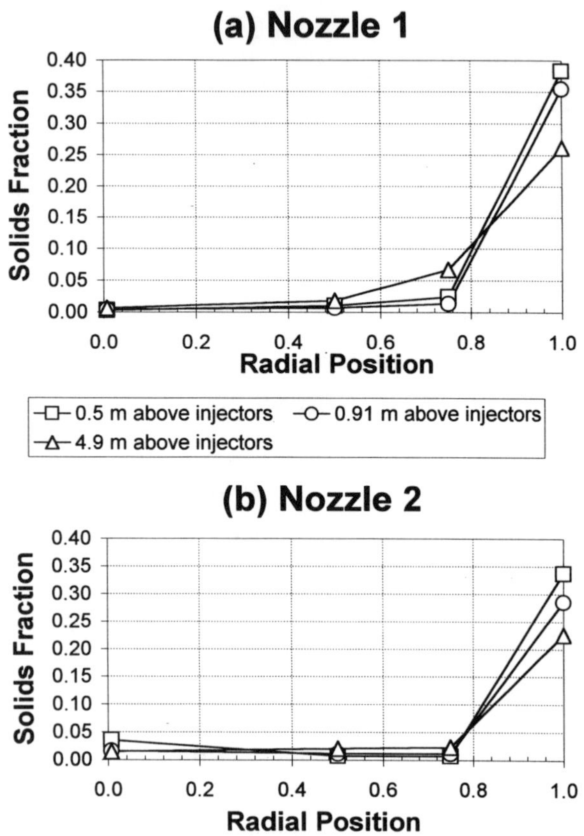

Figure 3: Radial solids fraction profile for the two different feed nozzles

CONCLUSIONS

Hydrodynamic measurements were made in a 0.45-m riser with liquid feed injection using two different nozzles. Qualitatively different catalyst mass flux, particle velocity and local catalyst density profiles were seen for the two nozzles. Feed from Nozzle 1, which yields a conical spray, penetrates to the center of the riser, forcing catalyst to the wall. Feed from Nozzle 2, which yields a flat spray, does not penetrate all the way to the center of the riser. In this case minima in the catalyst density and mass flux are seen off-center in the riser. Although after 11 L/D the profiles using both nozzles start to become similar, indicating that the effect of feed injection is beginning to dissipate, fully developed flow is still not reached. From these results it is apparent that the method of feed injection into risers is important and that the potential exists to tailor the catalyst hydrodynamic profiles in the feed injection zone.

LITERATURE CITED

1. **Monceaux, L., M. Azzi, Y. Molodtsof, and J.F. Large**, "Overall and Local Characterization of Flow Regimes in a Circulating Fluidized Bed," in "Circulating Fluidized Bed Technology," P. Basu, ed., Pergamon, New York, pp. 185-191 (1986).
2. **Bader, R., J. Findlay, and T.M. Knowlton**, "Gas-Solid Flow Patterns in a 30.5 cm Diameter CFB," in "CFB Technology II," P. Basu and J.F. Large, eds., Pergamon, New York, pp.123-127 (1988).
3. **Aguillon, J., K. Shakourzadeh, and P. Guigon**, "Comparitive Study of Non-Isokinetic Sampling Probes for Solids Flux Measurement in Circulating Fluidized Beds," *Powder Technology,* **83** (1), pp. 79-84 (Apr.1995).
4. **Kan, Z., J. Militzer, F. Hamdullahpur, and K. Shakourzadeh**, " Measurement of Particle Velocities in a Cyclone," in "Fluidization VIII," J.F. Large and C. Laguérie, eds., Engineering Foundation, New York, pp. 611-620 (1996).
5. **Louge, M. and M. Opie**, " Measurements of the Effective Dielectric Permittivity of Suspensions," *Powder Technology,* **62** (1), pp. 85-94 (Apr.1990).
6. **Lischer, D.J. and M. Louge**, "Optical Fiber Measurements of Particle Concentration in Dense Suspensions: Calibration and Simulation," *Applied Optics,* **31** (24), pp. 5106-5113 (Aug.1992).

DEM Simulation of Fluidized Beds of Polyolefin Synthesis

Yasunobu Kaneko*, Takeo Shiojima* and Masayuki Horio
Department of Chemical Engineering, Tokyo University of Agriculture and Technology, Koganei, Tokyo 184-8588 Japan
*Production Technology Center, Idemitsu Petrochemical Co., Ltd., Ichihara, Chiba 299-0193 Japan

Temperature behavior of particles and gas in a fluidized bed for polyolefin synthesis reactor was numerically analyzed based on the Discrete Element Method (DEM) with the soft sphere assumption. Simulation was performed by our numerical model (Kaneko et al. [1]) recently developed by incorporating the energy balance and the reaction rate to the code SAFIRE ver. 1 of Mikami, Kamiya and Horio [2].

To investigate the conditions of the hot spot formation, influence of the distributor types on bed temperature distribution was studied in more detail. Average and individual particle temperatures were examined. When gas was not uniformly supplied, rather stable particle swirls were formed in the bed and resulted in the hot spot. When the perforated plate was used as the distributor, the particle temperature tended to converge to a much higher value.

A fluidized bed type reactor has been widely employed for manufacturing polyolefins, especially polypropylene (PP) and polyethylene (PE), by using Ziegler-Natta catalysts of high activity and selectivity.

Since olefin polymerization is exothermic in nature, the temperature of polymer particles tends to rise and sometimes exceed the melting point of the polymer due to inadequate removal of the heat of polymerization from the growing polymer particle. In poor cooling conditions, polymer particles melt and form lumps which may then cause defluidization. Accordingly, the importance of heat removal from PO reactor has been recognized from the industrial view point. Several recent papers have reported modeling studies to analyze heat and mass transfer phenomena around a polymer particle by using the multigrain model [3,4,5]. However, these studies were on kinetic aspects and the behavior of polymerization reactors must be studied in consideration of particle and fluid dynamics in the reactor.

In this respect Kaneko et al. [1] applied recently a numerical simulation with DEM (Discrete Element Method), which was first applied to fluidized beds byTsuji et al. [6], to analyze the dynamic behavior of particle and gas temperatures in fluidized bed polyolefin reactors. It was found that the particle swirls induced by the bubble motion became hot spots. They also showed that the particle mixing condition in the bed is of great importance in hot spot formation. In the present study the conditions of the hot spot formation are investigated in more detail than our previous work [1], average and individual particle temperature histories were examined.

COMPUTATION CONDITIONS

Governing equations for gas phase and particle motion used in this study are the same as those given by our previous study [1] as summarized in Appendix. Detailed computation strategies are also the same as those of the previous paper. Here, it should be noted that both the direct particle-to-particle collision heat transfer and the radiative heat transfer were neglected for simplicity. Accordingly, the results include

Table 1 Computation conditions

Particles		
Number of particles,	n	14000
Particle diameter,	d_p	1.0×10^{-3} m
Restitution coefficient,	e	0.9
Friction coefficient,	μ	0.3
Spring constant,	k	800 N/m
Bed		
Bed size		0.154×0.3825 m (for n=14000)
Types of distributor		porous plate, perforated plate
Number of fluid cells		41×105 (for n=14000)
Time step		1.346×10^{-5} s

Table 2 Parameters for computation

Particles	density,	ρ_p	kg/m^3	717
	specific heat,	$c_{p,p}$	J/kg/K	2.30×10^3
	initial temperature,	$T_{p,0}$	K	363
Gas	density,	ρ_g	kg/m^3	20.44
	viscosity,	μ_g	Pa s	1.0×10^{-5}
	thermal conductivity,	λ_g	J/m/s/K	2.91×10^{-2}
	specific heat,	$c_{p,g}$	J/kg/K	1.80×10^3
	inlet gas temperature,	$T_{g,i}$	K	293
	pressure,	P	MPa	2.1
Reaction	Preexponential factor,	k_{p0}	kg/s/Pa/kg-cat	8.948×10^{-6}
	activation energy,	E	J/mol	10585
	catalyst weight,	w_c	kg/m^3	4.2×10^{-9}
	heat of polymerization,	$(-\Delta H_r)$	J/kg	3.77×10^3
u_{mf}	Predicted (Wen-Yu)		m/s	0.101
	Estimated by simulation		m/s	0.112

additional temperature rise due to these assumptions, but the effect of mixing on the temperature nonuniformity can be exaggerated.

Table 1 and 2 show computation conditions and physical parameters, respectively, used for calculation of ethylene polymerization. Physical properties are those taken at 363 K and 2.1 MPa for PE. A personal computer (Fujitsu Ltd., FMV-6266T6) was used for calculation.

RESULTS AND DISCUSSION

Figure 1 shows particle temperature and velocity distributions for three cases, where initial conditions

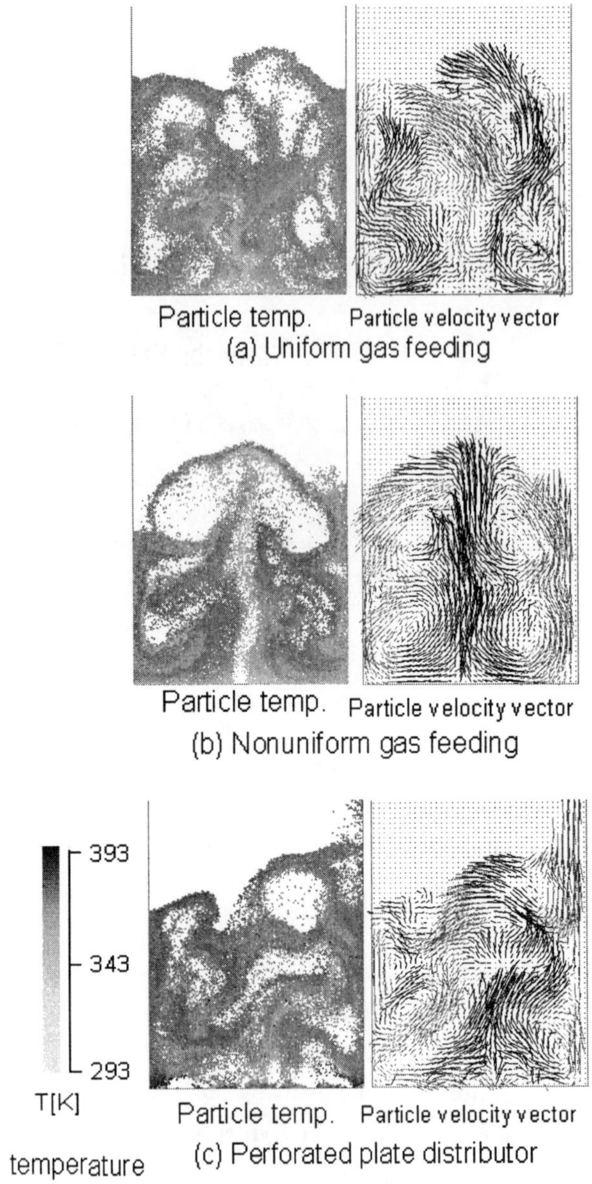

Figure 1 Particle temperatures and particle velocity vectors (number of particles=14000, PE reaction, $u_0=3u_{mf}$)

including the superficial gas velocity and the catalyst weight in a particle were all same. In the case of Figure 1(a) the superficial gas velocity u_y was three times as large as u_{mf} for all bottom cells, i.e. uniform gas feeding. In the case of Figure 1(b) u_y was set at 15.67 u_{mf} for three center cells and for the other 38 side cells at 2 u_{mf}. In the case of Figure 1(c) perforated plate was used as the distributor having 6

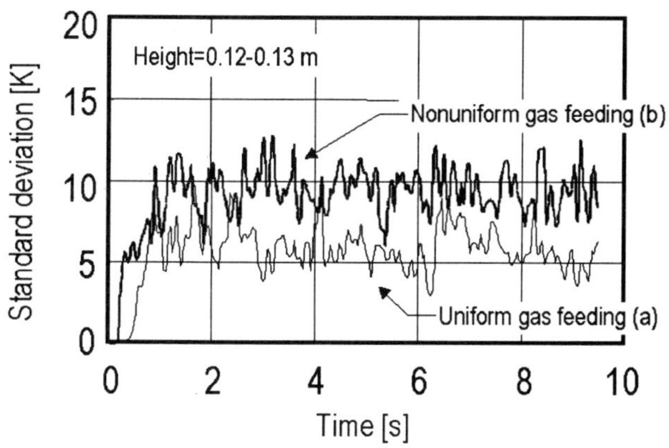

Figure 2 Standard deviations in case(a) and (b)

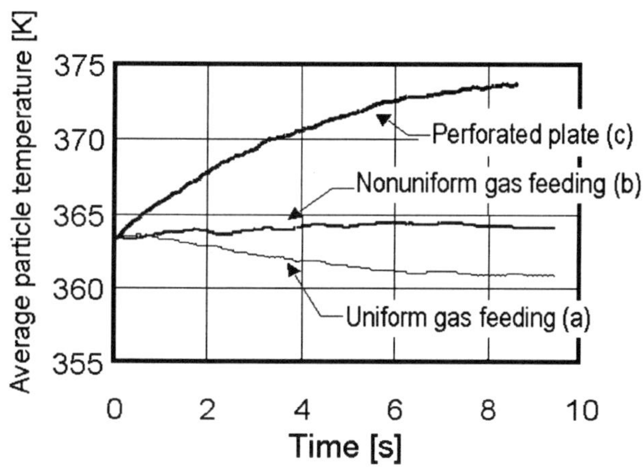

Figure 3 Histories of averaged particle temperature (number of particles=14000, PE reaction, $u_0=3u_{mf}$)

nozzles. In comparison with Figure 1(a) and Figure 1(b), the nonuniform gas feeding was found to create rather stable particle circulation eddies in the bed which caused the hot spot formation. To quantify the hot spot formation, the standard deviations around the radially averaged temperature profile were shown in Figure 2. The deviations were 5 K for (a) and 9 K for (b). In the case of the perforated plate distributor (Figure 1(c)) high temperature regions appeared throughout the bed in spite of the same random particle motion as Figure 1(a).

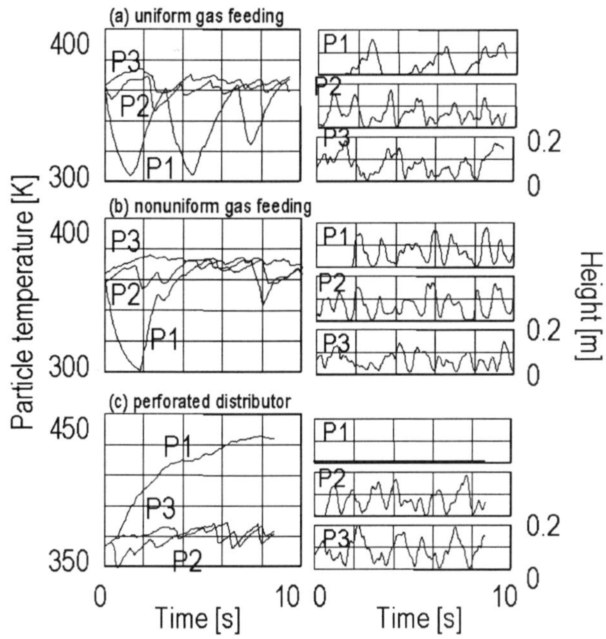

Figure 4 History of temperature and height of particle

To make the above point clearer, let us examine the histories of average particle temperature for the above three cases shown in Figure 3. As can be clearly seen in Figure 3, in cases of (a) and (b) the average temperatures approached rather quickly to their steady state values. On the other hand, in case of (c) the average particle temperature increased and finally approached to much higher steady state temperature than those in the former two cases. Figure 4 shows the temperature and height of three randomly chosen particles P1, P2 and P3 as a function of time. It can be found in Figure 4(a) and (b) that particle temperature was stable except when particles were cooled by moving down to the bottom zone. However, in Figure 4(c) the temperature of particle P1 continued to increase while it stayed in the bottom zone. This obviously implies that particle P1 was captured in the dead zone of higher temperature as can be seen Figure 1(c).

Figure 5 shows the snap shots of particles and their temperature for the perforated plate distributor.

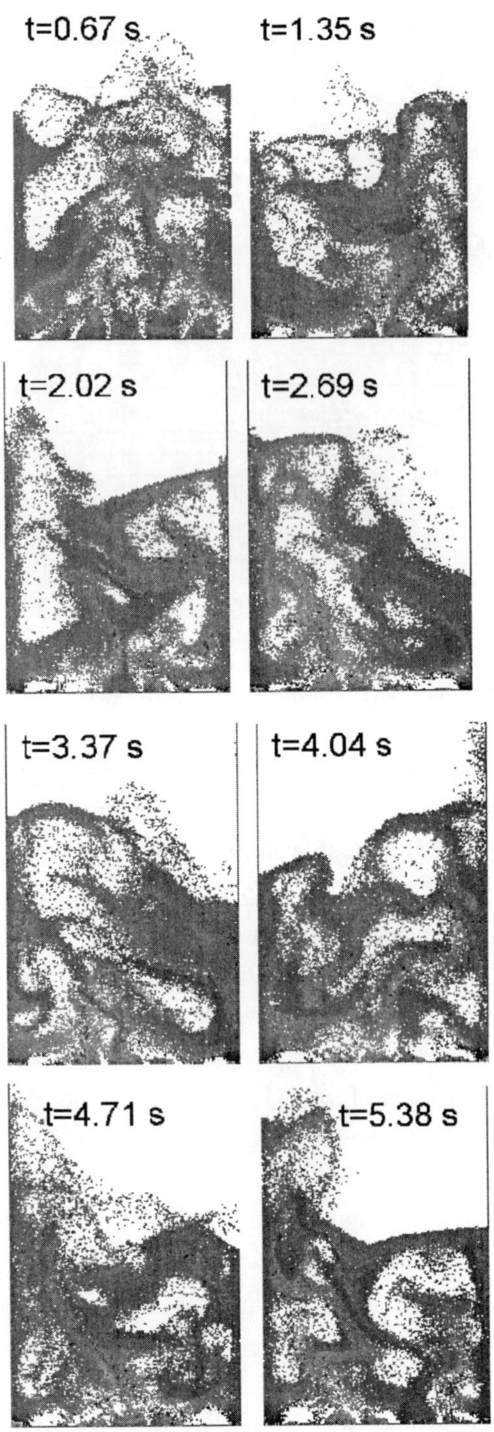

Figure 5 Snap shots of particles and with their temperatures (perforated plate distributor : 6 holes : $u_0=3u_{mf}$)

These results indicate that the random mixing of particles as well as the dead zone-less operation are important to prevent the hot spot formation, especially in cases of high heat of reactions.

CONCLUSIONS

By using the DEM simulation code SAFIRE [2] developed based on the soft sphere interaction during particle collision, the temperature behavior of particles and gas in fluidized-bed olefin polymerization was investigated with a special focus on the hot spot formation. It was found that the enhancement of randomness of particle mixing all over the bed was necessary to protect the hot spot formation.

NOTATION

$c_{p,g}$ =gas heat capacity, $J \cdot kg^{-1}K^{-1}$
$c_{p,p}$ =specific heat of solid, $J \cdot kg^{-1}K^{-1}$
d_p =particle diameter, m
E =activation energy, $J \cdot mol^{-1}$
f_i =fluid-particle interaction force acting on a fluid cell, N
g =gravity acceleration, $m \cdot s^{-2}$
h_p =heat transfer coefficient, $J \cdot K^{-1}m^{-2}s^{-1}$
$(-\Delta H_r)$ =heat of polymerization, $J \cdot kg^{-1}$
k_{p0} =preexponential factor, $kg \cdot s^{-1}Pa^{-1}kg\text{-}cat^{-1}$
Nu =Nusselt number, -
p_r =pressure, Pa
Pr =Prandtl number, -
q_p =convective heat flux, $J \cdot m^{-2}s^{-1}$
q_r =radiative heat flux, $J \cdot m^{-2}s^{-1}$
Q_g =heat exchanged between the particle and its external gas, $J \cdot kg^{-1}m^{-3}$
R =gas constant, $J \cdot mol^{-1}K^{-1}$
Re_p =Reynolds number, -
R_p = polymerization rate, $kg \cdot s^{-1}$
S_p =surface area of a particle, m^{-2}
t =time, s
T_g =gas temperature, K
T_p =particle temperature, K

u =gas velocity, $m \cdot s^{-1}$
v =particle velocity, $m \cdot s^{-1}$
V_p =volume of a particle, m^3
w_c =weight of catalyst in a particle, kg
x =horizontal coordinate, m
ε =voidage, -
λ_g =thermal conductivity of gas, $J \cdot K^{-1}m^{-1}s^{-1}$
μ_g =gas viscosity, Pa·s
ρ_g =gas density, $kg \cdot m^{-3}$
ρ_p =particle density, $kg \cdot m^{-3}$
σ =Stefan-Boltzmann constant, $W \cdot m^{-2}K^{-4}$

LITERATURE CITED

1. **Kaneko,Y., Shiojima,T. and Horio,M.,** "DEM simulation of fluidized beds for gas-phase olefin polymerization," *Chem.Eng.Sci.*, in press.
2. **Mikami,T., Kamiya,H and Horio,M.,** "Numerical simulation of cohesive powder behavior in a fluidized bed," *Chem.Eng.Sci.*, **53**, pp.1927-1940 (1998).
3. **Froyd,S., Choi,K.Y., Taylor,T.W. and Ray,W.H.,** "Polymerization of olefins through heterogeneous catalysis. Ⅲ .Polymer particle modeling with an analysis of intraparticle heat and mass transfer effects," *J.Appl.Polym.Sci.*, **32**, pp.2935-2960 (1986)
4. **Hutchinson,R.A. and Ray,W.H.,** "Polymerization of olefins through heterogeneous catalysis.Ⅷ.Monomer sorption effects," *J.Appl.Polym.Sci.*, **41**, pp.51-81 (1990)
5. **McKenna,T.F., Dupuy,J. and Spitz,R.,** "Modeling of transfer phenomena on heterogeneous Ziegler catalysts : Differences between theory and experiment in olefin polymerization (an introduction)" *J.Appl.Polym.Sci.*, **57**, pp.371-384 (1995)
6. **Tsuji,Y., Kawaguchi,T. and Tanaka,T.,** "Discrete particle simulation of two dimensional fluidized bed," *Powder Technol.*, **77**, pp.79-87 (1993)
7. **Ranz,W.E.,** "Friction and transfer coefficients for single particles and packed beds," *Chem.Eng.Prog.*, **48**, pp.247-253 (1952)

APPENDIX

Governing equations for particle motion and equation of gas continuity are the same as those given by Kaneko et al. [1]. The following equations were used for:

Gas momentum:

$$\rho_g \cdot \frac{\partial(\varepsilon \cdot u_i)}{\partial t} + \rho_g \cdot \frac{\partial(\varepsilon \cdot u_i \cdot u_j)}{\partial x_j} = -\varepsilon \cdot \frac{\partial p_i}{\partial x_i} - \varepsilon \cdot f_i + \varepsilon \cdot \rho_g \cdot g \tag{1}$$

Gas energy balance:

$$\frac{\partial(\varepsilon \cdot T_g)}{\partial t} + \frac{\partial(\varepsilon \cdot u_i \cdot T_g)}{\partial x_i} = \frac{Q_g}{\rho_g \cdot c_{p,g}} \tag{2}$$

where f_i is the fluid-particle interaction force, $c_{p,g}$ is the heat capacity of gas and T_g is the gas temperature. Here, radiation and direct particle-to-particle heat transfer were neglected first for simplification and second for creating larger temperature difference. Particle-to-particle direct heat transfer during collision requires the investigation on viscous damping by the thin fluid film between colliding particles, which should need substantial work in the future. As already discussed by Kaneko et al. [1], this simplification can exaggerate the effect of poor mixing on hot spot formation and therefore serves for the present objective to search for good design / operation strategies. To examine the extent of the effect of radiative heat transfer, radiative heat flux q_r is estimated by

$$q_r \equiv \sigma \cdot \left(T_i^4 - T_j^4\right) \tag{3}$$

where i and j are numbers of particles seeing each other. Suppose the temperature T_i can be as high as

373 K and T_j can be less than 363 K, we have q_r=113 J・m^{-2}s^{-1}. On the other hand, the convective heat flux q_p for minimum fluidization condition with $T_i - T_g$ =10 K is as high as 4000 J・m^{-2}s^{-1} from Ranz-Marshall [7] correlation, therefore neglecting q_r can be allowable. Of course, if T_i exceeds 400 K and T_j is about 300 K, q_r exceeds 1000 J・m^{-2}s^{-1} and cannot be neglected. Accordingly, the temperature rise in particle phase predicted by a model with these assumptions should be a little bit larger than the reality.

In Equation (2) Q_g is the heat transfer rate between a particle and gas in an unit volume expressed

$$Q_g = \langle h_p \cdot (T_p - T_g) \rangle \cdot 6 \cdot (1-\varepsilon)/d_p \quad (4)$$

where, h_p is the convective heat transfer coefficient between a particle and gas, d_p is the particle diameter and T_p is the temperature of a particle in the fluid cell. Heat transfer coefficient is determine by (Ranz-Marshall correlation [7]):

$$Nu = 2.0 + 0.6 \cdot \Pr^{1/3} \cdot \mathrm{Re}_p^{\ 1/2} \quad (5)$$

where,

$$Nu = h_p \cdot d_p / \lambda_g$$

$$\Pr = c_{p,g} \cdot \mu_g / \lambda_g \quad (6)$$

$$\mathrm{Re}_p = \rho_g \cdot d_p \cdot |u - v| / \mu_g$$

and, λ_g is the thermal conductivity of the gas, $c_{p,g}$ is the heat capacity of the gas, μ_g is the gas viscosity and $|u-v|$ is the gas-to-particle relative velocity.

The following was used for energy balance for a particle:

$$V_p \cdot c_{p,p} \cdot \rho_p \cdot \frac{dT_p}{dt} = R_p \cdot (-\Delta H_r) - \{h_p \cdot (T_p - T_g) \cdot S_p\} \quad (7)$$

where V_p is the volume of a particle, $c_{p,p}$ is the specific heat of solid, $(-\Delta H_r)$ is the heat of polymerization and S_p is the external surface area of a particle.

The reaction rate R_p is given by

$$R_p = k_{p0} \cdot \exp(-E/R/T_p) \cdot w_c \cdot p_r \quad (8)$$

where k_{p0} is the preexponential factor, E is the activation energy, w_c is the weight of catalyst in a particle and p_r is the monomer gas pressure.

Polymer Particle Mixing and Segregation in a Gas Phase Olefin Polymerization Reactor

Ju Yong Kim and Kyu Yong Choi
Department of Chemical Engineering, University of Maryland, College Park, MD 20742

In a gas phase fluidized bed olefin polymerization reactor with high activity transition metal catalysts, it is generally assumed that polymer particles are well mixed and almost isothermal reaction conditions prevail. However, it is often observed in some industrial fluidized bed polymerization processes that polymer particles in a fluidized bed reactor are often imperfectly backmixed. For example, the particle size distribution of polymers withdrawn from the bottom part of the fluidized bed is not same as that in the upper part of the reactor. In this paper, a multi-compartment model is presented to model the particle mixing and segregation phenomena in a gas phase olefin polymerization reactor. In this model, the particle transfer rate between the bulk emulsion phase and the wake phase is assumed to be particle size dependent. The model simulation results are presented for a seed bed and a reactive polymerization reactor systems to illustrate the feasibility of the proposed mixing-segregation model.

INTRODUCTION

Fluidized bed reactors are widely used in the polymer industry for the polymerization of α-olefins with transition metal catalysts such as Ziegler-Natta catalysts, chromium oxide catalysts, and supported metallocene catalysts. The fluidized bed polymerization processes are used to manufacture high density polyethylene (HDPE), low density polyethylene (LDPE), linear low density polyethylene (LLDPE), polypropylene (PP), ethylene-propylene copolymer and other ethylene-α-olefin copolymers. One of the major advantages of the fluidized bed reactor process is that solid polymer particles are intensely mixed and that the reaction heat can be effectively removed by the fast flowing fluidizing gas. The fluidized bed polyolefin process is currently a dominant industrial gas phase polyolefin process technology.

In a fluidized bed reactor system, a small amount of high activity catalyst particles (~ 30 μm) is supplied to the reactor. The polymerization occurs at the active sites in the solid catalyst and the catalyst particle injected to the reactor is quickly encapsulated by the growing polymers to a size of c.a. 300-1,000 μm. Fully grown polymer particles are withdrawn from the bottom portion of the reactor continuously or intermittently while keeping the bed level approximately constant. Since the gas velocity is high to maintain the particle fluidization, the monomer conversion per pass is quite low (< 5%) and a large amount of unreacted gas is cooled, compressed, and recycled back to the reactor for additional reaction and heat removal. As a result, the overall monomer conversion approaches near 100%.

It should be noted that the feed catalyst particles are not always of uniform size but have a certain size distribution. Since the size of a polymer particle produced in the reactor is determined by the particle's residence time (or reaction time) in the reactor, the polymer particles in a gas phase olefin polymerization reactor exhibit a broad particle size distribution. If catalyst deactivation occurs, the polymer particle size distribution becomes broader.

Although the polymer particles are generally assumed to be very well mixed in a fluidized bed polymerization reactor, particle segregation often occurs to some extent in industrial scale fluidized bed olefin polymerization reactors. For example, the size distribution of polymer particles removed from the bottom of the reactor may not be identical with the polymer particle size distribution in the reactor. If the control of polymer particle size distribution is important for post-reaction particle treatment, or if the polymer properties vary with size, understanding the non-uniform mixing of polymer particles in the reactor would be important. Sometimes a low fluidizing gas velocity is desired to reduce the operating cost and in such a case it is important to understand the effect of gas velocity on particle mixing and segregation as well as heat removal effi-

In a gas phase olefin polymerization reactor, the polymer particles of different sizes are almost of the same density. Although density difference is considered as the most powerful factor causing segregation in a fluidized bed, the particle size difference is thought to be the primary cause for particle segregation in a gas phase polymerization reactor. Yet, little has been reported on the particle mixing and segregation in gas phase fluidized bed polymerization reactors.

In this paper, we shall present a multicompartment model to model the mixing and segregation phenomena in a gas phase olefin polymerization reactor.

REACTOR MODEL

A two-phase bubbling bed model is widely used to model the gas phase fluidized bed polymerization reactor [1]. In the bubbling bed model, the solid phase (polymer phase) is assumed to be perfectly backmixed and the steady state polymer particle size can be calculated using the concept of particle residence time distribution. Population balance models have also been developed to predict the steady state polymer particle size distribution [2-4]. However, little has been reported on the nonuniformity of particle size distribution in a fluidized bed polymerization reactor.

It has been known that in an idealized case, solid particles are carried upwards in the wake and the drift of the bubbles [5,6]. As a gas bubble rises through the bubbling fluidized bed, the exchange of solid particles occurs between the wake and the surrounding bulk phases. The wake material dragged upward by the rising bubble is splashed onto the surface of the bed. Bubbles carry a wake of solids roughly equal to one third their own volume from bottom to top and so cause overall circulation [7]. Although the particle segregation phenomena have been studied extensively for binary mixture of particles of different densities or different sizes, little has been reported on the quantitative modeling of the particle segregation for a continuous size distribution.

To model the mixing and segregation phenomena in a fluidized bed gas phase polymerization reactor, we shall consider a multicompartment model as depicted in Figure 1. In this model, the entire reactor is divided into N virtual compartment units and each compartment units consists of a dense phase (emulsion phase) compartment and a lean phase (bubble/wake phase) compartment. The relative size or volume of

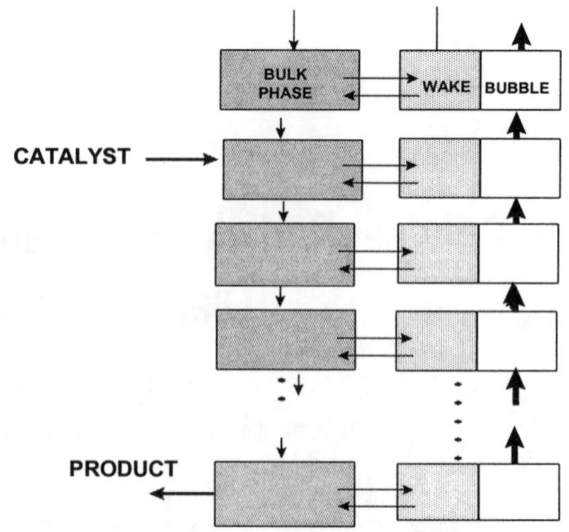

Figure 1 Multicompartment model.

each compartment in a unit can be approximated by the volume fraction of bubbles.

The following assumptions are made: (1) Bubbles are of uniform size and the total bubble holdup in the reactor is constant; (2) Solid particle rises only through the bubble phase (wake phase) and sink through the bulk emulsion phase only; (3) The voidage in the wake phase is the same as in the bulk emulsion phase; (4) Downward flow or movement of particles in the emulsion phase is not affected by particle size; (5) Particle elutriation is negligible. In this formulation, a separate bubble phase is not considered in the model. Instead, it is assumed that the bubble phase carries a certain fixed amount of solids in the wake.

In our model, we propose that the particle exchange between the emulsion phase and the wake phase is represented as

$$\mathbf{r} = k(r_p)P_e(r_p) - k'(r_p)P_w(r_p) \tag{1}$$

where k and k' are the particle size dependent particle transfer rate constants and P_e and P_w correspond to the particle concentrations in the emulsion phase and the wake phase, respectively. In this formulation k' represents the particle spillage from the wake phase to the bulk emulsion phase. This formulation is a departure from the model proposed by Gibilaro and Rowe [7] where the particle segregation is due to the preferential downfall of the larger/denser particles through the temporarily disturbed region behind the gas bubble.

Using the method described in [2], we can derive the steady state particle population balance equation for the n-th compartment unit as follows:

Emulsion phase:

$$\frac{dP_e(r_p,n)}{dr_p} = \frac{u_s}{h\mathbf{R}}\left\{P_e(r_p,n-1) - P_e(r_p,n)\right\} - \frac{A_w}{A_e\mathbf{R}}\left\{k(r_p)P_e(r_p,n) - k'(r_p)P_w(r_p,n)\right\} + \frac{5P_e(r_p,n)}{r_p} \quad (2)$$

where $\mathbf{R}$ is the particle growth rate ($= dr_p / dt$). For $r_p = r_c$ (catalyst particle radius), $q_{cn} = A_e h\mathbf{R}P_e(r_p,n)$ where q_{cn} is the catalyst injection rate. u_s is the downward solid flow rate in the emulsion phase and A_e and A_w represent the cross sectional areas of the emulsion and the wake phases. To calculate the rate of particle generation or growth, a polymer particle growth model (e.g., multigrain solid core model) [2] is solved with the polymerization kinetic model equations.

Wake phase:

$$\frac{dP_w(r_p,n)}{dr_p} = \frac{u_b}{h\mathbf{R}}\left\{P_w(r_p,n+1) - P_w(r_p,n)\right\} - \frac{1}{\mathbf{R}}\left\{k(r_p)P_e(r_p,n) - k'(r_p)P_w(r_p,n)\right\} + \frac{5P_w(r_p,n)}{r_p} \quad (3)$$

The particle population balances in the top compartment unit are represented as follows:

Emulsion phase:

$$\frac{dP_e(r_p,1)}{dr_p} = \frac{1}{h\mathbf{R}}\left\{\frac{A_w}{A_e}u_b P_w(r_p,1) - u_s P_e(r_p,1)\right\} - \frac{A_w}{A_e\mathbf{R}}\left\{k(r_p)P_e(r_p,1) - k'(r_p)P_w(r_p,1)\right\} + \frac{5P_e(r_p,1)}{r_p} \quad (4)$$

Wake phase:

$$\frac{dP_w(r_p,1)}{dr_p} = \frac{u_b}{h\mathbf{R}}\left\{P_w(r_p,2) - P_w(r_p,1)\right\} - \frac{1}{\mathbf{R}}\left\{k(r_p)P_e(r_p,1) - k'(r_p)P_w(r_p,1)\right\} + \frac{5P_w(r_p,1)}{r_p} \quad (5)$$

The particle population balances in the bottom compartment ($n = N$) are represented as follows:

Emulsion phase:

$$\frac{dP_e(r_p,N)}{dr_p} = \frac{1}{h\mathbf{R}}\left\{u_s P_e(r_p,N-1) - \frac{Q}{A_e}P_e(r_p,N)\right\} - \frac{A_w}{A_e\mathbf{R}}\left\{k(r_p)P_e(r_p,N) - k'(r_p)P_w(r_p,N)\right\} + \frac{5P_e(r_p,N)}{r_p} \quad (6)$$

Wake phase:

$$\frac{dP_w(r_p,N)}{dr_p} = -\frac{u_b P_w(r_p,N)}{h\mathbf{R}} - \frac{1}{\mathbf{R}}\left\{k(r_p)P_e(r_p,N) - k'(r_p)P_w(r_p,N)\right\} + \frac{5P_w(r_p,N)}{r_p} \quad (7)$$

The polymer withdrawal rate (Q) can be calculated from the overall solid (polymer), catalyst, and active site balances at steady state and it is expressed as

$$Q = \frac{1}{2}\left\{\sqrt{(Wk_d)^2 + 4Wq_c R_{p0}} - Wk_d\right\} \quad (8)$$

where R_{p0} is the initial polymerization rate (i.e., without deactivation), W is the bed weight, and k_d is the deactivation rate constant.

In the proposed model, it is assumed that particle segregation occurs due to size-dependent particle transfer between the emulsion and the wake phases. Then the particle transfer rate constants need to be available. Chan and Knowlton [8] show that the particle size distribution of the entrained solids approaches that of the whole bed at high gas velocities while the distribution shifts to smaller sizes at low gas velocities. The entrained solids are mainly the particles from the wake thrown into the freeboard above the dense phase with the burst of gas bubbles. Although the mechanisms of elutriation and the particle transfer between the emulsion and bubble phases are not same, we shall assume that the size-dependent mass transfer rate parameters is approximately represented by such a model. Thus, the following particle transfer rate constants for a diameter of d_p are used in our model simulations [9].

$$k(d_p) = A\rho_p\left(1 - \frac{u_t}{u_0}\right)^2$$
$$k'(d_p) = k(2d_{pm} - d_p) \quad (9)$$

where A is an adjustable parameter, d_{pm} is the median value of particle sizes in a mixture. Also, the following equations are used to estimate the terminal velocity of a particle.

$$u_t = \left[\frac{18}{\left(d_p^*\right)^2} + \frac{2.335 - 1.744\phi_s}{\left(d_p^*\right)^{0.5}} \right]^{-1} \left[\frac{\mu\left(\rho_p - \rho_g\right)g}{\rho_g^2} \right]^{1/3} \tag{10}$$

$$d_p^* = d_p \left[\frac{\rho_g\left(\rho_p - \rho_g\right)g}{\mu^2} \right]^{1/3} \tag{11}$$

ϕ_s is the sphericity and μ is the gas viscosity.

RESULTS AND DISCUSSION

Let us first consider a seed-bed fluidized bed reactor. The seed-bed reactor refers to the reactor charged with dead or inactive polymer particles. In a fluidized bed polymerization process, the reactor is started up by first charging the reactor with polymer particles of known properties and known particle size distribution. The steady fluidization is established by supplying an inert gas or a mixture of inert gas and monomer such as ethylene. After the reactor temperature is brought to a certain target temperature, the injection of high activity catalyst is started.

We assume that initially the polymer particles are perfectly mixed in the reactor. The solid particles move upward with fast rising bubbles, fall and mixed with other solid particles in the bulk emulsion phase. We assume that the bulk density of the polymer particles is constant (This assumption may not be valid if partial melting or sintering of polymer particles occurs to cause particle agglomeration). Figure 2 shows the polymer particle size distribution curves in the top and bottom compartments for the fluidizing gas velocity of 0.2 m/sec and 0.6 m/sec. The symbols represent the industrial test data (scaled). Notice that the model gives a reasonable prediction of the particle size distribution. It is seen that at high gas velocity there is no discernible difference in the particle size distributions for the top and the bottom compartments. At low gas velocity, the difference becomes more apparent.

The multicompartment model has been applied to an ethylene homopolymerization process with supported Ziegler-Natta type catalyst whose particle size

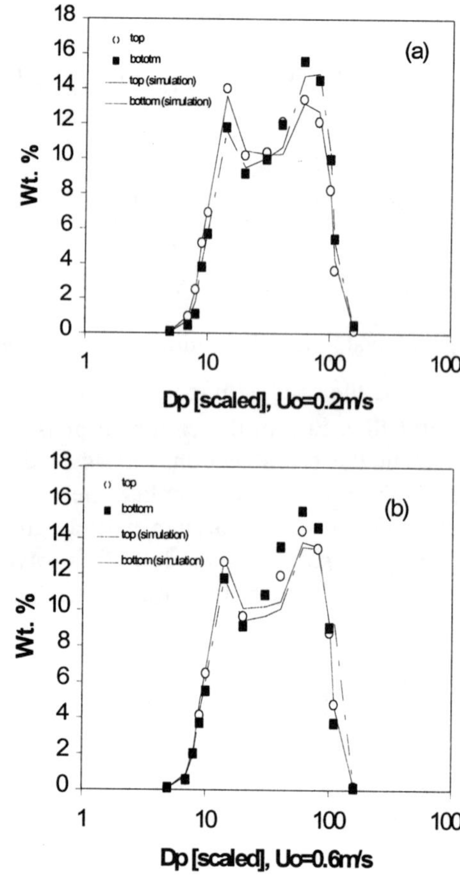

Figure 2 Particle size distribution in a seed bed reactor.

distribution is shown in Figure 3. Figures 4(a) and 4(b) show the polymer particle size distributions in the top and the bottom compartments at steady state for different fluidizing gas velocities for a polymerization reactor. Notice that there is a significant difference in the polymer particle size distributions at the top and the bottom portion of the reactor. The concentration of large particles is higher in the bottom of the reactor. It is because the rate of spillage of large particles from the wake is higher than that for small particles. If the polymer particles of different sizes or different residence times have different properties, the properties of the polymer particles withdrawn from the reactor will be the average properties.

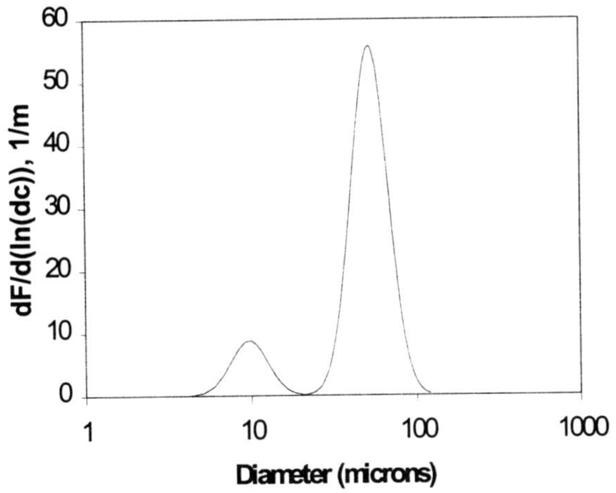

Figure 3 Input catalyst particle size distribution.

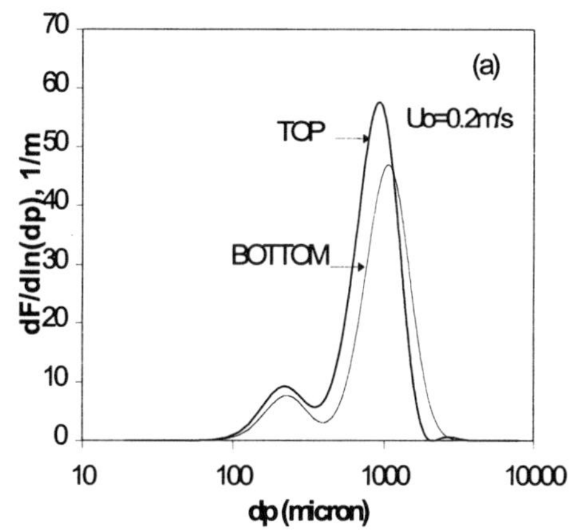

Figure 4(a) Polymer particle size distribution.

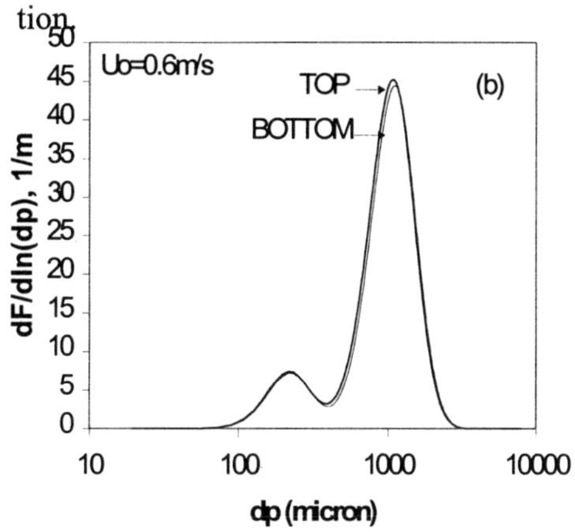

Figure4(b) Polymer particle size distribution.

CONCLUDING REMARKS

In this paper, we proposed a multicompartment model to model the particle size distribution in a gas phase olefin polymerization reactor. In this model, the rate of particle transfer from the bulk emulsion phase to the wake phase and the rate of particle spillage from the wake phase to the bulk emulsion phase are assumed to be particle size dependent. Since there is a lack of experimental data on the size-dependent particle transfer in a fluidized bed reactor, we adopted the structure of the rate parameter proposed for the particle elutriation process. Although this assumption may not be fully validated, the proposed model shows that it can be used to predict the demixing phenomena in a fluidized bed reactor. The model simulation results indicate that the fluidizing gas velocity has a strong influence on the segregation of polymer particles. Using the model proposed, we can investigate the effect of various reactor operating conditions on the polymer particle size distribution in a gas phase olefin polymerization process.

NOTATION

A = cross sectional area, cm^2

h = height of a bed segment, cm

k = size dependent particle transfer rate constant (into wake), 1/s

k' = size dependent particle spillage rate constant (from wake), 1/s

k_d = catalytic site deactivation rate constant, 1/s

P = particle size distribution function

Q = polymer withdrawal rate, g/s

q_c = catalyst injection rate, g/s

R_{p0} = initial polymerization rate, g/g-cat.s

r_p = polymer particle radius, cm

r_c = catalyst particle radius, cm

u_b = bubble rise velocity, cm/s

u_0 = superficial inlet gas velocity, cm/s

u_s = solid descending velocity in bulk phase, cm/s

$\mathbf{R}$ = particle growth rate, cm/s

W = total bed weight, g

Greek letters

μ = gas viscosity, g.cm/s

ρ_p = particle density, g/cm^3

ρ_g = gas density, g/cm^3

ϕ = particle sphericity

Subscripts

e = bulk (emulsion phase)
w = wake phase

LITERATURE CITED

1. **Choi, K.Y., and W.H. Ray**, "The Dynamic Behavior of a Fluidized Bed Rreactor for Solid Catalyzed Gas Phase Olefin Polymerization Reactor," *Chem. Eng. Sci.*, **40**(12), 2261-2279 (1985).
2. **Choi, K.Y., X. Zhao, and S. Tang**, "Polulation Balance Modeling for a Continuous Gas Phase Olefin Polymerization Reactor," *J. Appl. Polym. Sci.*, **53**, 1589-1594 (1994).
3. **Zacca, J.J., J.A. Debling, and W.H. Ray**, "Reactor Residence Time Effects on the Multistage Polymerization of Olefins -I. Basic Principles and Illustrative Examples, Polypropylene," *Chem. Eng. Sci.*, **51**, 4859-4886 (1996).
4. **Hatzantonis, H., A. Goulas and C. Kiparissides**, "A Comprehensive Model for the Prediction of Particle Size Distribution in Catalyzed Olefin Polymerization Fluidized-Bed Reactors," *Chem. Eng. Sci.*, **53**(18), 3251 (1998).
5. **Baeyens, J., and D. Geldart**, in "Gas Fluidization Technology," ed. D. Geldart, 97, Wiley, New York, 1986.
6. **Hoffmann, A.C., L.P.B.M. Janssen and J. Prins**, "Particle Segregation in Fluidised Binary Mixtures," *Chem. Eng. Sci.*, **48**(9), 1583-1592 (1993).
7. **Gibilaro, L.G., and P.N. Rowe**, "A Model for a Segregating Gas Fluidised Bed,"*Chem. Eng. Sci.*, **29**, 1403-1412 (1974).
8. **Chan, L.H., and T.M. Knowlton**, "The Effect of Pressure on Entrainment from Bubbling Gas-Fluidized Bed," in "Fluidization IV", D. Kunii and R. Toei eds., 283-290, Engineering Foundation, 1983.
9. **Kunii, D., and Levenspiel, O.**, "Fluidization Engineering", 2nd ed.,Butterworth-Heinemann, Boston, MA, 211-235 (1991).

Solid State Shear Extrusion (SSSE) Pulverization of Rubber Granules

Ecevit Bilgili, Hamid Arastoopour, and Barry Bernstein
Illinois Institute of Technology, Department of Chemical & Environmental Engineering, 10 West 33rd Street Chicago, Illinois

A single screw extruder was used to pulverize vulcanized rubber granules in the solid state into fine particles under high shear and normal forces. This process, known as Solid State Shear Extrusion (SSSE), is non-cryogenic and has the capability of producing rubber particles having diameters ranging from 40 microns to about 1500 microns. Because the rubber granules were partially de-vulcanized under high shear forces, the produced particles were very suitable for recycling. During pulverization, a significant part of the strain energy which has been stored in the rubber at high shear and normal forces is released in the form of heat. This causes the temperature of the produced powder to increase. Rubber particles at high temperature and under normal forces can initiate an oxidation process and formation of undesired agglomerates. At a very high temperature, the particles may even degrade. This means the rate of heat transfer is one of the most critical parameters in controlling the sizes of the produced particles. Thus, this study focuses not only on the fundamentals of pulverization and processing of the particles but also on experimental and numerical analyses of heat transfer and the material temperature distribution throughout the process.

Recycling of waste rubber poses a significant environmental problem since it cannot be melt-processed due to its inherent cross-linked structure. The Solid State Shear Extrusion (SSSE) pulverization process, which is a non-cryogenic mechano-chemical process, can convert the waste rubber into powder in a continuous manner.

Enikolopian [1] and his co-workers conceptualized the use of an extruder to pulverize plastics under compressive shear. Wolfson [2] et al. investigated the possible scale-up problems of the process such as frictional heating of polymers during pulverization. Arastoopour [3] et al. focused on the effects of SSSE process variables on the quality of the produced rubber powder and showed that there are optimum operating conditions at which powder quality is enhanced.

The aim of this study is to experimentally and computationally show the importance of the heat transfer rate for efficient pulverization of rubber granules. We also investigate possible heat removal systems such as internal cooling of a hollow screw.

EXPERIMENTAL STUDY

The extruder used to pulverize rubber granules is a lab-scale single screw extruder manufactured by C.W. Brabender that has a barrel with a diameter of 1.9 cm, and a length to diameter ratio of 25 (see Figure 1). It consists of three zones and each zone has a cooling jacket in contact with the barrel wall. Pulverization takes place mostly in the third zone of the extruder where the material is subjected to sudden cooling and compressive shear [4]. The screw used in the experiments is a solid screw with no internal cooling. Large rectangular slabs of rubber samples were obtained from Coopers Tire and Rubber Company and shredded before each experiment into small granules of 3-6 mm in diameter.

In the first set of experiments, we examined the feasibility of pulverization of the rubber granules using the SSSE process. Rubber granules at room temperature were fed via a hopper at a rate (hfr) of 11.6 gr/min. The 1st, 2nd and 3rd zone barrel wall temperatures (T_{b1}, T_{b2} and T_{b3}, respectively) were set at 40°C. The rotational rate of the screw (ω) was set at 80 RPM. The experiment was completed in 5 minutes to prevent the heating of the screw and the material due to heat dissipation associated with the pulverization process. In other words, we tried to minimize the effects of non-isothermal and unsteady regime on the particle size distribution. The experiment was repeated several times to minimize the effect of possible non-uniformity in the feeding process. The particle size distribution of the produced particles was measured by sieves and a Microtrack particle size analyzer. To understand the effect of reprocessing, the produced powder, which is the one pass SSSE product (1SSSE), was put into the extruder and pulverized a second (2SSSE) time and then a third time (3SSSE).

The particle size distribution (PSD) and mean diameter (D_m) of the powder after each processing are shown in Figure 2. The particle sizes varied from 40 microns to about 1500 microns. Compared to the 1SSSE case, the powder quality was better in 2SSSE and 3SSSE due to the reduction in initial feed size. However, the 3SSSE powder was not better than the 2SSSE powder because about 10% of the fine particles (< 150 μm) were lost during reprocessing of the 2SSSE powder via agglomeration: reprocessing causes the breakage of very big particles and agglomeration of the finer ones and hence makes the PSD narrower.

The agglomeration of rubber material could be due to an oxidation reaction in the solid state at elevated temperatures and high normal forces. To minimize the agglomeration of produced rubber particles, we tried to keep the material temperature in the third zone below the characteristic oxidation temperatures of rubber. The extruder was run for no more than 5 minutes to minimize the temperature rise in the extruder. However, we obtained some agglomeration of the fine rubber particles, particularly during the 2SSSE and 3SSSE runs, probably due to high material temperature at high shear and normal forces.

In another set of experiments, we investigated the effects of non-isothermal regime on the powder quality. The extruder was operated at the same conditions as described before. However, the experiments were continued for a long time, which allowed us to study the development of the temperature profile in the extruder. Seven different temperatures may be measured in our experiments: barrel wall temperatures T_{b1}, T_{b2}, T_{b3}, material temperatures T_{m1}, T_{m2}, T_{m3} and screw end temperature T_s. J type thermocouples were aligned flush with the material at the end of each zone to measure the material temperatures at the material-barrel wall interfaces. Other thermocouples were placed at the middle of each zone and measured the barrel wall temperatures in contact with the cooling jacket. A thermocouple was placed on the screw end and rotated with the screw to prevent pseudo-frictional heating. To understand the effect of cooling, two cases were studied. In the first case, neither air cooling (in the 1st and 2nd zones) nor water cooling (in the 3rd zone) was provided and the temperatures at various locations were recorded (see Figure 3a). In the second case, both air and water cooling in the above-mentioned zones were supplied and the pertinent temperatures were recorded (see Figure 3b).

The effect of heat generation due to pulverization can be seen at all temperatures (see Figure 3a). Since there was no cooling or heating, the increase in temperatures can be attributed only to equivalent work, i.e., frictional heat generated at the screw and material interface and heat dissipation caused by the conversion of stored energy into heat when the microcracks formed and pulverization began. In 14 minutes, the third zone barrel wall temperature rose from 40°C to about 90°C. Very big agglomerates were exiting the extruder. We believe that, due to very sharp temperature gradients in the material, oxidative degradation occurred in some portions of the material, which caused the agglomeration of the particles. We could not measure these high local temperatures since any thermocouple inserted in a moving solid would be distorted at high shear and would produce erroneous temperatures. The increase in the temperatures of other barrel walls was not very significant. This indicates that pulverization takes place in the third zone of the extruder where compressive shear is exerted on the material in the narrow channel between the barrel and the screw. The increase in all other temperatures could be due to the back conduction of heat from the third zone or frictional heat generated at the interface and screw due to screw rotation and slip between the screw and material. The increase in temperatures was different in the cooling experiment (see Figure 3b). T_{b3} was kept almost constant at 47°C due to water cooling. The other barrel wall temperatures did not increase appreciably either. On the other hand, the screw end temperature rose to 62°C in 30 minutes. This was expected since the screw was not hollow and had no means of transferring heat to the surroundings except through axial conduction. T_s almost reached a steady-state after 30 minutes. In the cooling case, we could produce powder for 30 minutes, but some agglomerates were formed near the end of the experiment when the screw temperature reached its maximum.

We believe that the energy stored in the material is converted into heat during pulverization. The energy is given to the material by the screw, and the torque developed in the screw and the rate of rotation of the screw determine the total energy and, in turn, the amount of heat dissipation. The reprocessing experiments were repeated for 20 minutes and the screw end temperature was recorded. The torque (τ) developed and the screw end temperature were directly related (see Figure 4). As shown in Figure 4, torque was higher in 1SSSE compared to 2SSSE and 3SSSE. We had very good control over torque in 3SSSE since the interstitial void was minimized due to the presence of narrower size distribution. The heat generation rate was also higher (higher screw end temperature) in 1SSSE than in 2SSSE and 3SSSE. This experimental result showed a direct correlation between the torque developed due to the frictional forces and the heat generation rate.

COMPUTATIONAL STUDY OF HEAT TRANSFER IN SSSE PULVERIZATION PROCESS

To study the heat transfer in the 3rd zone of the extruder, a mathematical model was developed based on the following assumptions: the motion of the particles in the 3rd zone is assumed to be plug-flow ($V_\theta \approx 0$) with a solids loading of 11.6 gr/min; the granules and particles act like an incompressible solid, and complex multidimensional deformation field of the rubber under high shear and normal forces is not considered; the inlet is assumed to be isothermal (at 25°C); the initial temperature for the screw (solid or hollow), the material (rubber) and the barrel wall are considered to be 25°C; the temperature and flux is continuous at the solid-solid interfaces; Newton's Law of Cooling at solid-cooling medium surfaces is considered and all model parameters are considered to be constant. The geometry of the third zone of the extruder with a hollow screw is shown in Figure 5. The parameter values pertinent to heat transfer are listed in Table 1.

A heat transfer model based on the differential energy balances for the screw, material moving at a constant velocity (with a continuous heat source due to pulverization), and barrel wall was used to solve for the temperature field using a FTCS (Forward Time Centered Space) Explicit Finite Difference method on a 51x51 mesh in cylindrical coordinates. The heat generated due to pulverization is expressed as:

$$Q_{gen} = \frac{0.99\,\tau\,\omega}{\pi\left(R_3^{\,2} - R_2^{\,2}\right)L}$$

We assumed that 99% of the total energy supplied to the material by the screw is converted into heat during pulverization. The heat dissipation is uniformly distributed along the third zone of the extruder in our model. We have experimental data on the torque developed as a function of time and the rotational rate of the screw (80 RPM). This information was used in Q_{gen} term.

The simulation results for a solid screw with no inner cooling and a hollow screw with inner cooling after a processing time of 600 s were plotted as shown in Figure 6(a) and 6(b) respectively. The screw temperature (Region 1), the material temperature (Region 2) and the barrel wall temperature (Region 3) at each spatial point ($r^* = r/L$ and $z^* = z/L$) approach steady-state after 600 s.

Figure 6(a) shows that there are sharp temperature gradients in the rubber due to its very low thermal conductivity. The temperature gradients in the solid screw and the barrel wall are smaller than those in the rubber since the solid screw and the barrel wall have a thermal conductivity an order of magnitude higher than the one of the rubber. At the exit of the third zone (z=L), the axial temperature gradient is nearly zero because ambient air cannot efficiently remove heat via natural convection. The maximum barrel wall temperature predicted by the model is around 130°C, which may cause both agglomeration of the fine particles as well as degradation of the rubber. Heat can be removed very effectively from the third zone when a hollow screw, which is cooled by chilled water, is used in the extruder (see Figure 6b). The sharp gradients still exist in the rubber although the screw, material and barrel wall temperatures are very low compared to the solid screw case. To minimize the temperature gradients, a cooling medium in direct contact with the material, such as a cold inert gas, should be used. This can be achieved by fluidization of the particles after pulverization using a cold gas.

There is a discrepancy between our simulation results (see Figure 6a) and the experimental results (see Figure 3b). We could measure material temperatures only at the interfaces. These temperatures may actually be much lower than the bulk rubber temperatures. The experiments could not reveal possible radial temperature gradients, which can only be predicted by modeling at that stage. The interface temperatures predicted by the model are much higher than those measured experimentally. This is mainly due to the presence of a heat sink that is not considered in simulations and the assumption of 99% conversion of stored energy into heat. SSSE pulverization of rubber results not only in size reduction, but also, in the breakage of some weak chemical bonds such as S-S disulfide bonds (de-vulcanization). This chemical change is endothermic and acts as a heat sink. The estimation of the heat sink due to de-vulcanization necessitates further research on the mechano-chemical aspects of the process. There is no calorimetric measurement known to estimate the heat dissipation under high mechanical stresses. An improved model should take these considerations into account to estimate the temperature profile in the extruder more accurately. The present model guides the design of the cooling system and enables us to study possible radial temperature gradients in the extruder.

CONCLUSION

Our experimental and computational studies showed that the SSSE pulverization process has the capability to produce rubber particles with an efficient cooling system. The SSSE process is non-isothermal and unsteady in nature and the desired steady-state may not be reached unless a highly efficient cooling system is used. Experimental data must be gathered and evaluated considering the non-isothermal and unsteady nature of the process.

NOTATION

C_P: heat capacity
D: equivalent diameter of rubber particles
h: convective heat transfer coefficient of a fluid
k: thermal conductivity
L: length of the third zone of the extruder
r: radial coordinate axis
R: location of an interface
T: temperature
V: velocity
z: axial coordinate axis
ρ: density

ACKNOWLEDGEMENT

The authors would like to thank the NSF Fluid, Particulate and Hydraulic Systems Division for financial support of this work through Grant # CTS-9629650.

LITERATURE CITED

1. **Enikolopian, N.S.,** "Some Aspects of Chemistry and Physics of Plastic Flow," *Pure and Appl. Chem.*, **57** (11), pp. 1707-1711 (1985).
2. **Wolfson, S.A., Nikolskii, V.G.,** "Powder Extrusion: Fundamentals and Different Applications," *Polym. Eng. Sci.*, **37** (8), pp. 1294-1300 (1997).
3. **Arastoopour, H., Bilgili, E., Schocke, D. and Bernstein, B.,** "Pulverization of Rubber under High Shear and Normal Forces," *Proc. World Cong. Particle Technol. 3*, Brighton, U.K., Paper No. 127 (1998).
4. **Arastoopour, H.,** "Single Screw Extruder for Solid State Shear Extrusion Pulverization," U.S. Patent 8,101,468 (1998).

	Material Parameters
For Rubber	C_p=1884 J•kg^{-1}K^{-1}, k=0.173 W•m^{-1}K^{-1}, ρ=967 kg•m^{-3}
For Screw and Barrel	C_p=460 J•kg^{-1}K^{-1}, k=45.0 W•m^{-1}K^{-1}, ρ=7884 kg•m^{-3}
	Convective Heat Transfer Parameters
Solid Screw Case	h_{f1}, T_{f1} (no fluid); h_{f2}=300 W•m^{-2}K^{-1}, T_{f2}=25°C; h_a=10 W•m^{-2}K^{-1}, T_a=25°C
Hollow Screw Case	h_{f1}=h_{f2}=300 W•m^{-2}K^{-1}; T_{f1}=T_{f2}=0°C; h_a=10 W•m^{-2}K^{-1}, T_a=25°C

Table 1. The parameter values used in the heat transfer computations

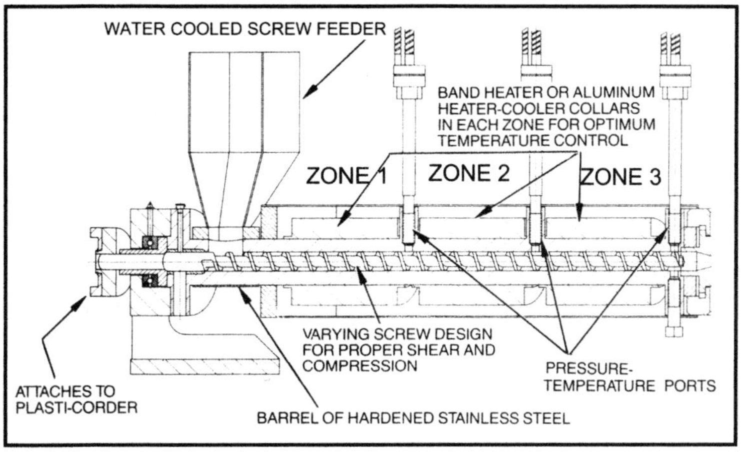

Figure 1. A schematic of the single screw extruder used in the pulverization of rubber

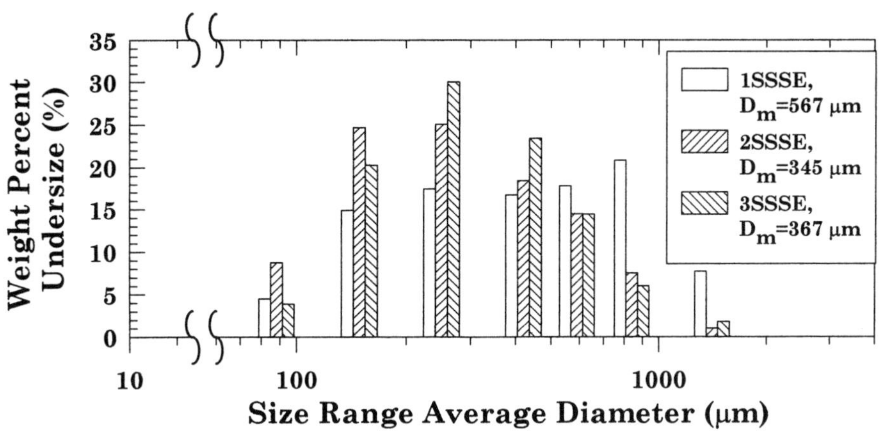

Figure 2. Effect of reprocessing on particle size distribution (PSD) at barrel wall temperatures of T_{b1}=T_{b2}=T_{b3}=40°C, screw rotational rate of 80 RPM and hopper feed rate of 11.6 gr/min

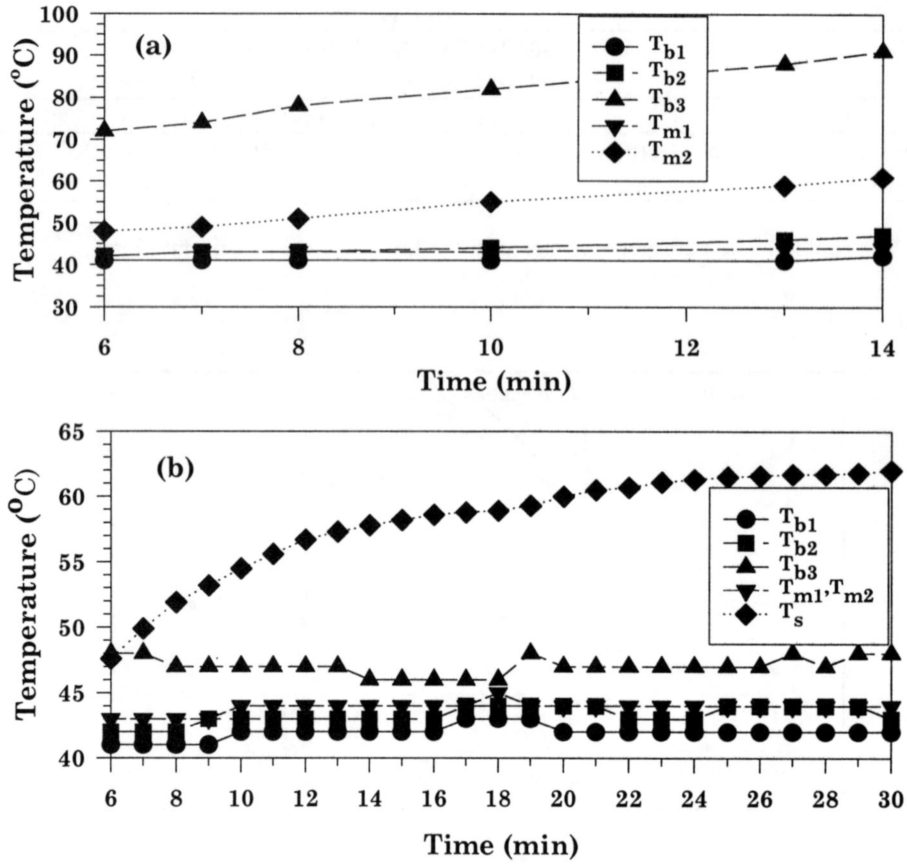

Figure 3. Temperature profile in the extruder as a function of time: (a) no air or water cooling, (b) air and water cooling

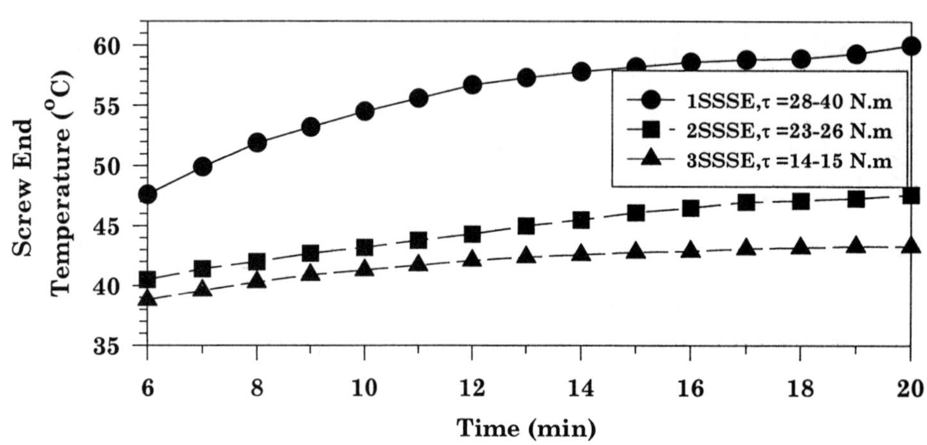

Figure 4. Unsteady behavior of screw end temperature in 1SSSE, 2SSSE and 3SSSE

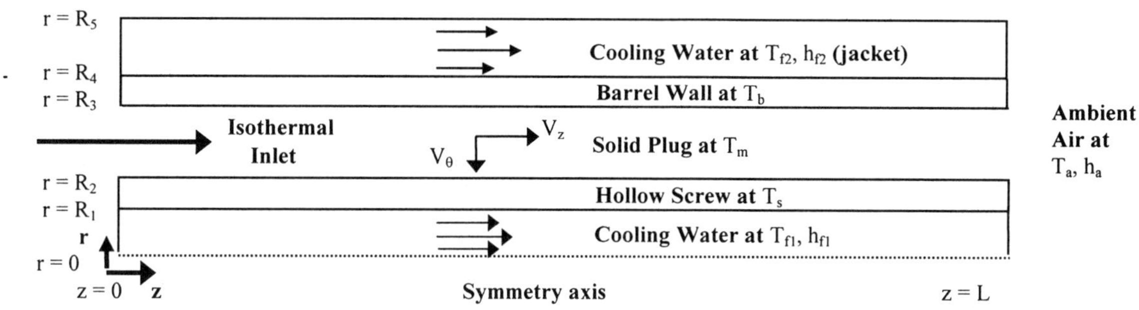

Figure 5. Model geometry: third zone of the single screw extruder with a hollow screw

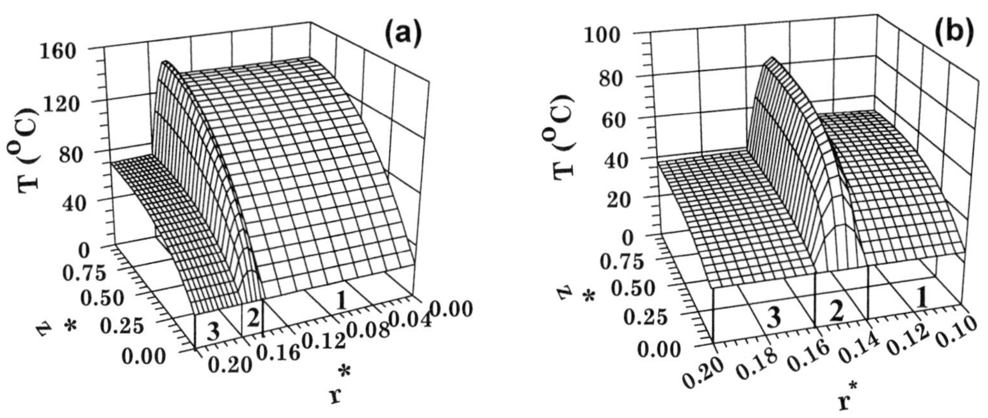

Figure 6. Temperature profile of the third zone of the extruder: (a) solid screw case, (b) hollow screw case

Characterization of Suspension Flows Using Particle Image Velocimetry (PIV)

O.S. Hasan, M.M. Alvarez, F.J. Muzzio and H.M. Buettner
Department of Chemical and Biochemical Engineering, Rutgers University, Piscataway NJ 08855

Particle image velocimetry (PIV) techniques were used for the simultaneous measurement of solid phase velocity and concentration fields of suspensions agitated in a stirred tank geometry. This study demonstrates that PIV can resolve solid phase velocity fields for concentrated suspensions of large particles. Although PIV typically provides only velocity information, algorithms have been developed to derive solid concentration measurements from the captured PIV images. Solid glass beads of various sizes were suspended in a mixture of glycerin and water in a stirred tank. Digital images of these suspensions were taken and particle concentration profiles were obtained from the images. Segregated regions within the suspension flow can be identified in both the suspended phase velocity and concentration fields.

The suspension of solid particles in a liquid medium occurs commonly in industrial practice, particularly for chemical reaction and mass transfer operations. For example, in pharmaceuticals manufacture, crystallization and dissolution operations both involve the use of suspended phases. In these situations, mixing plays a critical role in determining the physical properties of the products. Poor mixing during crystallization, for instance, may lead to broad crystal size distributions [1]. The optimization of suspension processes requires the assessment of suspension homogeneity. The homogeneity of solid suspensions is often studied by examining the solids concentration at a few points within the mixing system [2-5]. The velocity of the suspended phase is usually not measured. A more complete characterization of suspension mixing requires evaluating both the suspended phase concentration and velocity fields at a large number of points in the system.

In this work, the suspended phase velocity and concentration fields are simultaneously measured using particle image velocimetry (PIV). PIV is usually employed to rapidly measure velocity fields in single-phase fluid flows [6]. During a typical PIV measurement, a flow is seeded with a minute quantity of small (< 10 microns), neutrally-buoyant tracer particles that are assumed to follow the fluid flow closely. A pulsed laser light sheet is used to illuminate a two-dimensional plane of the flow and images of the light scattering are captured with a digital camera. Velocity fields are generated from these images using a computer algorithm.

This paper describes the use of PIV as a tool for assessment of suspension homogeneity. Only a few studies thus far have demonstrated the usefulness of PIV to characterize suspensions possessing a significant solids concentration [7-9]. Previous work has focused on development of the technique, rather than drawing conclusions about particular flow systems, and was largely performed in uncommon flow geometries. Here, PIV is used to characterize suspension flows in a typical stirred tank geometry. The concentration of the suspended particles used in this work was significantly higher than in typical PIV applications. Also, these particles were much larger than the tracer particles normally employed. Particles of this size were used for PIV in fluidized beds, however their utilization for PIV in stirred tanks has not yet been reported [10, 11]. In addition to the measurement of suspended phase velocity fields, spatial variations in suspended phase concentration were monitored using an image analysis algorithm developed in-house. The simultaneous measurement of the suspended phase concentration and

---

Rutgers University, Piscataway, New Jersey.

velocity field provides information necessary for assessing suspension homogeneity.

EXPERIMENTAL PROCEDURE

The mixing vessel used in the experiments reported here was a flat-bottomed, cylindrical glass tank of inner diameter T = 110 mm. A three-blade marine-type impeller of diameter D = 50 mm was installed, closely matching a ratio of T/D = 2, common for stirred tank geometries. The impeller clearance C = 50 mm was approximately equal to T/2. The working fluid was a mixture of glycerin (99.7% USP) and water (95%/5% v/v). The refractive index of this mixture closely matched that of glass, which was useful for improved suspension imaging. The working fluid was filled to a height H = T = 110 mm. The impeller and the impeller shaft were painted a matte black finish to minimize light reflection. To reduce optical distortion due to the curved surface of the mixing vessel, the vessel was placed in a rectangular glass tank and the space between the mixing vessel and the tank was filled with glycerin.

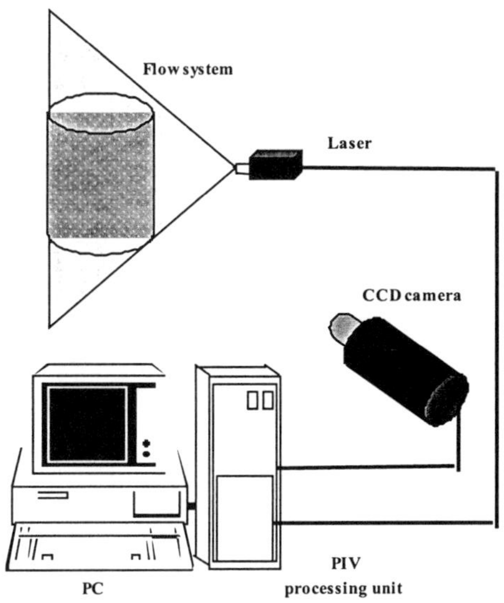

Figure 1. Schematic representation of PIV system.

The particles used in the experiments were Dragonite solid glass beads (Jaygo, Inc., Union, NJ) with a mean diameter of 600 μm and a size range of 490 μm - 700 μm. The specific gravity of the glass beads was 2.55, and the solids concentration in all experiments was 1.0% by volume. Experiments were carried out at impeller speeds from 600 to 1500 rpm, corresponding to impeller Reynolds numbers of 41.2 to 103 (laminar flow regime). At agitation speeds greater than 1500 rpm, vortex effects become significant and incorporated air bubbles interfere with the PIV measurements. For all cases, the agitation speed was sufficient to bring about complete suspension conditions according to the Zwietering criterion [12].

The concentration and velocity fields of the glass beads were measured with a PIV processing unit (Dantec Measurement Technology, Skovlunde, Denmark). The suspended phase was illuminated by a pulsed laser light source and gray-scale images of the illuminated beads were captured with a digital camera located perpendicularly to the illuminated plane. The flow illumination and image capturing system is depicted in Figure 1. From the captured images, suspended phase velocity and concentration fields were obtained by image analysis techniques. For a particular experimental run, twenty pairs of images of the flow were captured. For each image pair, a velocity field was derived using a cross-correlation computer algorithm that compared particle positions between the two images. In this manner, twenty velocity fields were constructed for a particular experimental run. These velocity fields were averaged to minimize random noise in the measurements.

Particle concentration information was also obtained by analysis of the captured images of the glass beads. An algorithm has been developed in-house for analysis of these images [13]. It is not possible to determine the actual number of particles contained in the plane of each image for two reasons. First, some particles may be only partially contained in the imaging plane. Second, the scattering of light off particles contained fully in the imaging plane may be highly variable. However, relative changes in particle concentration in various regions of the mixing vessel can be captured from the images. To obtain the suspended phase concentration field for a particular experimental run, the twenty captured images were analyzed by dividing each image into a number of interrogation areas and determining the number of pixels containing particle reflection in each area. The twenty concentration fields were averaged to reduce the effect of local particle concentration fluctuations.

RESULTS AND DISCUSSION

The suspended phase velocity field is shown in Figure 2 for agitation at a rate of 800 rpm. This plot represents the suspended phase velocity field in a vertical plane through the center of the mixing vessel. The radial and axial velocity components can be measured from this velocity field. The tangential velocity component can be obtained in a similar manner by measuring the suspension velocity field at a particular horizontal cross-section through the mixing vessel. Figure 2 reveals two circulatory flow structures within the cross-section, one above and one below the impeller. Similar flow structures have been observed at all agitation speeds used in this work. These structures correspond to toroid-shaped recirculation regions of regular motion that have been well characterized for single-phase fluid flow in stirred tanks operated in the laminar regime [14-16].

For single-phase fluid systems, it is well known that these toroid-shaped recirculation regions remain completely segregated from the rest of the flow domain, where chaotic motion is experienced by the fluid particles [17]. A fluid tracer injected inside one of these segregated regions may only leave that region by diffusion. While it has been shown that both bubbles and particles can remain trapped in the areas of regular motion, the presence of such segregated regions in suspension flows has not been documented before [18]. Here, we provide such evidence by means of the inspection of particle concentration fields.

Suspension concentration fields were constructed by the method described above. Figure 3 shows the average suspension concentration field for the same plane within the mixing vessel, again at 800 rpm. The abscissa corresponds to the radial position in the tank, where zero is the tank center, while the ordinate relates the axial position, where zero is the tank bottom. In this plot, the impeller is located at a vertical position of 50 mm and extends radially from the center of the tank to a radial position of approximately 25 mm. Areas of high solids concentration are colored white and areas of low concentration are shown as black. Intermediate concentration levels are given as shades of gray. Figure 3 depicts two significant areas in the concentration field of zero particle concentration. These areas, which appear as dark regions in the concentration fields, correspond to the torroid-shaped areas of regular motion mentioned previously. The results indicate that particles are not present in the recirculation regions and these regions are segregated with respect to the rest of the flow domain. The presence of these segregated regions in the flow has been observed even at high

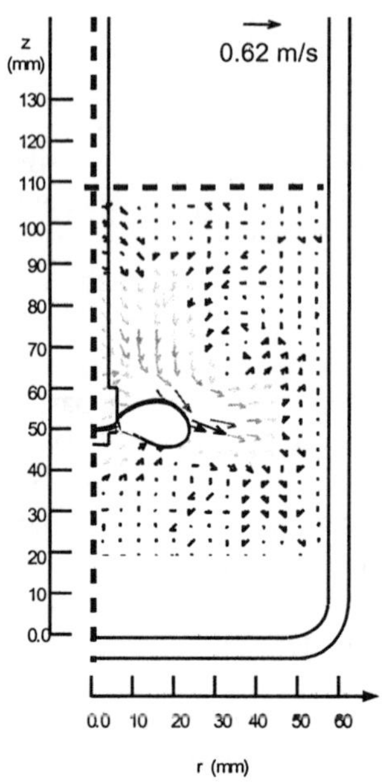

Figure 2. Suspended phase velocity field in a vertical plane through the center of the mixing vessel. Agitation speed was 800 rpm.

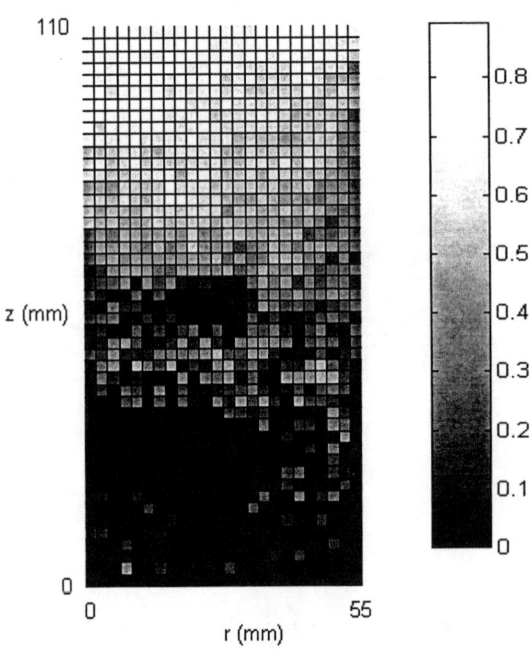

Figure 3. Suspended phase concentration field in a vertical plane through the center of the mixing vessel. Dark regions indicate low particle reflection. Agitation speed was 800 rpm.

impeller speeds, although their location and shape characteristics vary with impeller speed. This fact will be explored in detail in a future communication.

CONCLUSIONS

This paper discusses the extension of particle image velocimetry (PIV) techniques for characterization of suspension flows operated in the laminar regime. The PIV technique has been used for simultaneous measurement of suspended phase velocity and concentration fields in stirred tanks. Although PIV typically provides only velocity information, algorithms have been developed to derive solid concentration measurements from the captured PIV images. Measurement of the suspended phase concentration and velocity field provides information useful for assessing suspension homogeneity. Segregated regions within a suspension flow were identified through analysis of the suspended phase velocity and concentration fields. The techniques described here will be useful in investigating the effects of process parameters, such as agitation rate and system geometry, on suspension mixing performance.

SYMBOLS AND UNITS

C = clearance of impeller off the tank bottom, mm
D = impeller diameter, mm
H = height of liquid in mixing vessel, mm
r = radial position in vessel, mm
T = vessel diameter, mm
z = axial position in vessel, mm

LITERATURE CITED

1. **Tavare, N.S.,** "Industrial Crystallization: Process Simulation Analysis and Design," Plenum Press, New York (1995).
2. **Baldi, G., R. Conti, and A. Gianetto,** "Concentration profiles for solids suspended in a continuous agitated reactor," *AIChE J.*, **27**(6), p. 1017-1020 (1981).
3. **Bos, A.S. and J.J. Heerens,** "Light backscattering as a technique to measure solids particle size and concentration in suspension," *Chem. Eng. Commun.*, **16**, p. 301-311 (1982).
4. **Ayazi Shamlou, P. and E. Koutsakos,** "Solids suspension and distribution in liquids under turbulent agitation," *Chem. Engng. Sci.*, **44**(3), p. 529-542 (1989).
5. **Godfrey, J.C. and Z.M. Zhu,** "Measurement of particle-liquid profiles in agitated tanks," *AIChE Symp. Ser.*, **90**(299), p. 181-185 (1994).
6. **Adrian, R.J.,** "Particle-imaging techniques for experimental fluid dynamics," *Annu. Rev. Fluid Mech.*, **23,** p. 261-304 (1991).
7. **Zachos, A., M. Kaiser, and W. Merzkirch,** "PIV measurements in multiphase flow with nominally high concentration of the solid phase," *Exp. Fluids*, **20**(3), p. 229-231 (1996).
8. **Chen, R.C. and L.S. Fan,** "Particle image velocimetry for characterizing the flow structure in three-dimensional gas-liquid-solid fluidized beds," *Chem. Engng. Sci.*, **47**(13/14), p. 3615-3622 (1992).
9. **Farrell, P.V.,** "Particle sizing and velocity measurements using particle image velocimetry," in "Flucome '91 - Third Triennial International Symposium on Fluid Control Measurment and Visualization," ASME (1991).
10. **Chen, R.C., J. Reese, and L.S. Fan,** "Flow structure in a three-dimensional bubble column and three-phase fluidized bed," *AIChE J.*, **40**(7), p. 1093-1104 (1994).
11. **Reese, J., R.C. Chen, and L.S. Fan,** "Three-dimensional particle image velocimetry for use in three-phase fluidization systems," *Exp. Fluids*, **19**(6), p. 367-378 (1995).
12. **Zwietering, T.N.,** "Suspending of solid particles in liquid by agitators," *Chem. Engng. Sci.*, **8**(3/4), p. 244-253 (1958).
13. **Alvarez, M.M. and F.J. Muzzio,** "Determination of solid particle concentration profiles in stirred vessels using particle laser tomography," *Exp. Fluids (submitted)*, (1999).
14. **Fasano, J.B. and A. Bakker,** "Analysis of the flow field in agitated tanks using particle image velocimetry," *Mixing XIV*, Santa Barbara (1993).
15. **Lamberto, D.J., *et al.*,** "Using time-dependent RPM to enhance mixing in stirred vessels," *Chem. Engng. Sci.*, **51**(5), p. 733-741 (1996).
16. **Lamberto, D.J., M.M. Alvarez, and F.J. Muzzio,** "Experimental and computational investigation of the laminar flow structure in a stirred tank," *Chem. Engng. Sci. (accpeted for publication)*, (1999).
17. **Lamberto, D.J., M.M. Alvarez, and F.J. Muzzio,** "Computational analysis of regular and chaotic mixing in a stirred tank reactor," *Chem. Engng. Sci. (submitted)*, (1999).

18. **Lamberto, D.J.,** "Enhancing mixing in stirred tank reactors using time-dependent agitation rates," Chemical and Biochemical Engineering, Rutgers University: Piscataway, NJ (1997).

Continuous Measurement of Velocity and Direction of Particles Using the Lag-Time Method

Hiroyuki Hatano, Satoru Matsuda
Thermal Energy and Combustion Engineering Department, National Institute for Resources and Environment
Tsukuba 305-8569, Japan
Katsumi Tsuchiya
Department of Chemical Science and Technology, The University of Tokushima, Tokushima 770-8506, Japan

A new, simple algorithm is proposed for simultaneous measurement of the velocity and the trajectory of solid particles in unsteady gas-solid flow. It utilizes the time difference of the particle peaks between the main and reference signals obtained using a reflective optical fiber probe to calculate the particle velocity. The critical step in the algorithm is to determine the corresponding peaks for the identical particles; the closest peaks in the pair of signals are selected. This algorithm is applied for relatively low concentration of particles where the distance between particles is greater than twice the distance between the main and reference optical fibers.

The accuracy of the proposed algorithm is confirmed using a free-fall particle system. The particle velocity in a circulating fluidized bed (CFB) is also examined. The results show that this method can be used to continuous measurement the particle velocity and direction.

INTRODUCTION

In the operation of gas-solid flow processes, online monitoring of the behavior of solid particles provides useful information, since the solid particles affects reactions, heat transfer, and flow dynamics of the process. Many researchers have tried to develop methods of such particle monitoring. For instance, the strain gauge technique was applied to measure the bulk solids flow rate in a fluidized bed [1]. The hot wire method was applied to obtain the same information [6]. These methods, however, cannot be used to detect the time variation of the particle velocity.

As a real-time particle velocimetry, methods utilizing the optical fiber probe and the data processing system have been developed, including the direct measurement of time intervals between particle signals [4] and the spatial filtering method [2]. These methods are effective in measuring the velocity variation for particles in unidirectional motion, but otherwise not suitable for finding the direction of particles. To overcome this problem, different data processing techniques have been proposed, including the short integration-period cross correlation method [3]. Through this method, both the speed and direction of solid particles can be estimated simultaneously in a two-dimensional domain. This algorithmis not a realistic one to use it as an online monitoring of the particle flow, because of numerous calculation steps.

In this study, a simple algorithm for online monitoring of both the particle speed and direction is proposed. Besides the basic concept behind the algorithm, the results of measurements for two systems, a free-fall particle system and a circulating fluidized bed (CFB) system, are demonstrated.

DATA PROCESSING

Figure 1 shows a schematic of the tip arrangement of the optical fiber probe used in this study. The probe, being a reflective type, consists of three optical fibers aligned vertically, the center of which emits light. When particles are passing the probe (in the figure, vertically upward), two series of signals are obtained which signify the passage of the particles by distinctive peaks. Two peaks appearing on each channel correspond to the two ascending particles. The particle velocity, v_p, can be calculated from the distance between the light-receiving fibers, Δl_f, and the time lag, τ_1, as

$$v_p = \Delta l_f / \tau_1 \qquad (1)$$

Depending on the direction of the particles, the time lag could be positive or negative. The algorithm developed here determines not only the time difference directly but also the moving direction.

Peak Detection

To reduce the noise from the original signal, the method of moving average is adopted; in this study, five consecutive points are averaged. The increment in signal level between succeeding points is then calculated for the portions exceeding a prefixed threshold level. We designate the points at which the positive sign of the increment turns negative as peak candidates. These candidates are further screened and regarded as true peaks when the difference between the peak height and the base line of the signal exceeds another (higher) threshold. This procedure is schematically shown in Fig. 2.

Detection of Moving Direction

In the schematic illustration of the signals shown in Fig. 1. One peak is marked in the main channel, CH1, and corresponding to this peak on the positive and negative sides of CH2 are registerd. The closer peak was chosen as the corresponding peak when the following conditions are satisfied:

(1) The direction of particle motion aligns with that of the optical fibers.

(2) The particle velocity is constant until the particle has passed the optical fiber probe.

(3) The concentration of solid particles is dilute and the distance between particles is longer than twice the distance between the receiving optical fibers. That is,

$$2\Delta l_f < \Delta l_p \quad (2)$$

This criterion is reasoned as follows. As indicated in Fig. 1, the period required for a particle passing the distance Δl_p is calculated as

$$\tau_1 + \tau_2 = \Delta l_p / v_p \quad (3)$$

For the ascending particles,

$$\tau_1 = \Delta l_f / v_p, \quad \tau_2 = (\Delta l_p - \Delta l_f)/v_p \quad (4)$$

Condition (2) is then shown to be equivalent to $\tau_1 < \tau_2$, i.e.,

$$\tau_1 - \tau_2 = (2\Delta l_f - \Delta l_p)/v_p < 0 \quad (5)$$

For the descending particles, the same relationship can be obtained. Hence, the closest peaks between two channels are recognized as the footprints of the same particle when Condition (2) holds.

EXPERIMENTAL

The data processing scheme proposed above needs to be standardized via calibration. Two calibration methods are adopted using a rotating disk and free-fall particles, and a continuous measurement has been made in the riser of a circulating fluidized bed cold model.

Figure 3 shows the outline of the data processing system. The light signals received by the optical fibers are converted to electric signals by a photomultiplier, amplified up to a voltage range between 0 and 5 V by operational amplifiers, and stored in a digital data recorder at 20 kHz sampling rate for the most of the runs. The sampled data are processed offline after all the experiments have been made.

Calibration using Rotating Disk System

The rotating disk system, shown in Fig. 4, is used for the calibration and for the inspection of the applicability of the technique into the dense particle flow. On a black disk, white lines of 0.2-mm width are drawn radially. The optical fiber probe is set at a certain radial position, 33 mm from the center for this case, and the alignment of the light-receiving fiber tips is made onto the tangential line. The velocity calibration is then made based on the known angular velocity of the disk.

Figure 5 shows the relationship between the velocity of the white lines on the disk, which simulates the particle velocity v_p, and the measured reciprocal lag time determined based on the probe signal. The slope of the correlated line is Δl_f, which is calculated to be 0.37 mm.

Calibration with Free-Fall Particle System

In order to calibrate the particle velocity in dilute particle flow, free-fall velocity of 0.40-mm glass beads is measured using a system shown in Fig. 6. Particles flowing out of the funnel are straightened out by an obstructing plate to avoid the effect of initial uneven acceleration.

A typical pair of raw signals and the free-fall velocities of solid particles evaluated based on the obtained Δl_f are shown in Fig. 7. The particle velocities thus obtained are all (except a few) close to -1 m·s^{-1}, signifying a uniform downward flow of particles realized in the experiment. Figure 8 shows the averaged velocities obtained at different vertical locations, H below the damper plate (Fig. 6), compared to the corresponding values estimated from the analytical formula, $(2gH)^{1/2}$. The two sets of velocities are in reasonable agreement.

Circulating Fluidized Bed Riser

The test apparatus used in this study is a transparent CFB cold model and its schematic diagram is shown in Fig. 9. Both the riser and the downcomer were 60 mm in inner diameter and 4500 mm high. The optical fiber probes were placed at two locations, 700 and 3600 mm, from the distributor. Solids loading rate, G_S, was controlled by altering the aeration velocity through an H-valve which connects the riser and downcomer at the bottom. In this study, it was fixed at 10 kg·m^{-2}s^{-1}. Glass beads of 0.40-mm mean diameter were used as the fluidized particles; the bulk density is about 1530 kg·m^{-3} and the minimum fluidization velocity 0.20 m·s^{-1}. The superficial gas velocity, U_0, ranges from 2 to 6 m·s^{-1} and most of the data are obtained at 4 m·s^{-1}.

RESULTS AND DISCUSSION

Circulating Fluidized Bed Riser

Figures 10 and 11 show two sets of raw signals and the resulting velocity variations obtained in the central and the wall regions, respectively, at the upper section, h = 3600 mm. The corresponding results for the central region of the lower section, h = 700 mm, are shown in Fig. 12. In the central region of the upper section, ascending particle velocities were frequently observed and their values range from 4 to 5 m·s^{-1}. Descending particle velocities were less frequently observed and their absolute values were much lower than the ascending velocities. In the wall region, most particles had descending velocities, the magnitude of which ranged between0.5 to 1 m·s^{-1}; the ascending velocities of particles were measured as 1 m·s^{-1}. The particle velocity measured at the lower section ranged from 1 to 5 m·s^{-1} which was slightly lower on average than that at the upper section because the lower section was considered to be within the acceleration zone.

Data Processing Rate

The software in this study has been developed using the Turbo-C programming language. With this, the continuous measurement required for the online monitoring of the particle behavior can be realized in a microcomputer system (in this study, a PC486-DX2

66MHz is used). The present limitation is that the code requires 1-s computation time for a set of data sampled over 0.1 s to display the calculated average velocity on the monitor. This intermittency could, however, still be regarded as "continuous" for the particle monitoring in practical operation of systems such as CFBs.

Despite its physical soundness, the cross correlation method requires a longer calculation period. It is also often required to check the obtained value of the cross correlation function by eye-inspecting the original time series data. In the present method, the data reliability of the generated data can be inspected in the peak detection step by adjusting the parameters.

CONCLUSIONS

A new data processing techni ue is introduced to monitor the velocity and the moving direction of the solid particles. The following results are obtained:

1. By devising the software logic, simultaneous measurements of the particle velocity and the moving direction become possible.
2. The proposed method has shorter data processing period than the cross correlation method, and thus can be used as an online monitoring system of the particle behavior.
3. The algorithm for the peak detection and discrimination consists of very simple subroutine packages so that this program can be readily replaced with an improved algorithm for higher accuracy and shorter peak-detection period.

* This technique has been tested for one dimensional flows, however, it is also applicable to the two dimensional measurement by using the probe consisting of six reflective optical fibers. The relationship between the combination of reflective optical fibers and the velocity has been discussed in the previous papers [3,7].

NOTATION

g =	gravitational acceleration	$[\mathrm{m \cdot s^{-2}}]$
G_S =	solids loading rate	$[\mathrm{kg \cdot m^{-2} s^{-1}}]$
h =	distance above distributor	[m]
H =	distance traveled by a free-fall particle	[m]
U_0 =	superficial gas velocity	$[\mathrm{m \cdot s^{-1}}]$
v_p =	particle velocity	$[\mathrm{m \cdot s^{-1}}]$
Δl =	distance	[m]
τ =	time	[s]

Subscripts

1 forward
2 backward
f optical fiber
p particles

LITERATURE CITED

1) **Hirato, M.**, T. Miyamoto, J. Tamura, S. Yoshioka, N. Kiuchi, and Y. Ishibashi, *Fluidization Science and Technology*, pp. 91–99, Science Press (1982).
2) **Ishida, M.**, K. Ogawa, and T. Shirai, *Kagaku Kogaku Ronbunsyu*, **5**, 487–492 (1979).
3) **Ishida, M.**, T. Shirai, and A. Nishiwaki, *Powder Technol.*, **27**, 1–6 (1980).
4) **Ogawa, K.**, H, Sato, M. Ishida, and T. Shirai, *J. Chem. Eng. Japan*, **11**, 410–412 (1978).
5) **Oki, K.**, T. Akehata, and T. Shirai, *Kagaku Kogaku*, **37**, 965–967 (1973).
6) **Yamazaki**, M., and Y. Kato, *Kagaku Kogaku Ronbunsyu*, **5**, 155–161 (1979).
7) **Oki, K.**, M. Ishida, and T. Shirai, Fluidization III, pp. 421-426 , Engineering Foundation (1980)

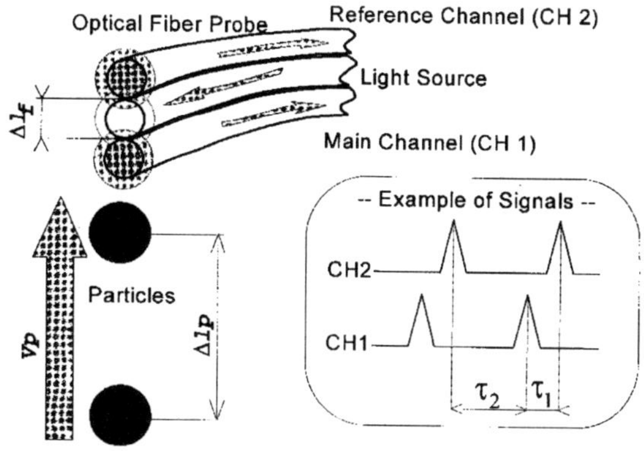

Fig.1 Schematics of the probe and the particle detecting principle.

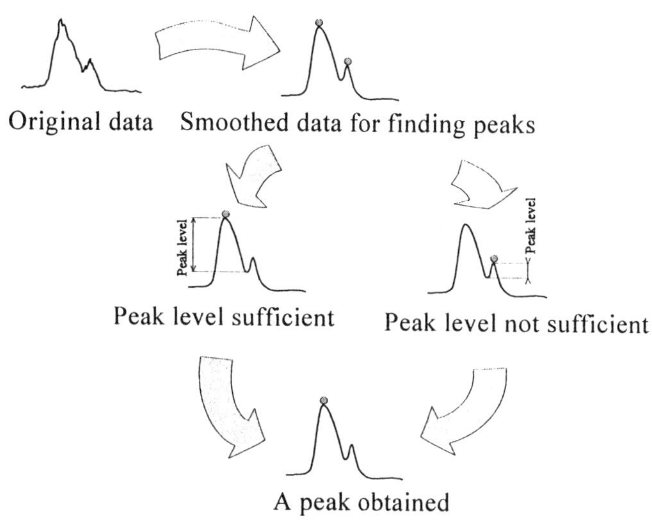

Fig.2 Procedure for peak detection.

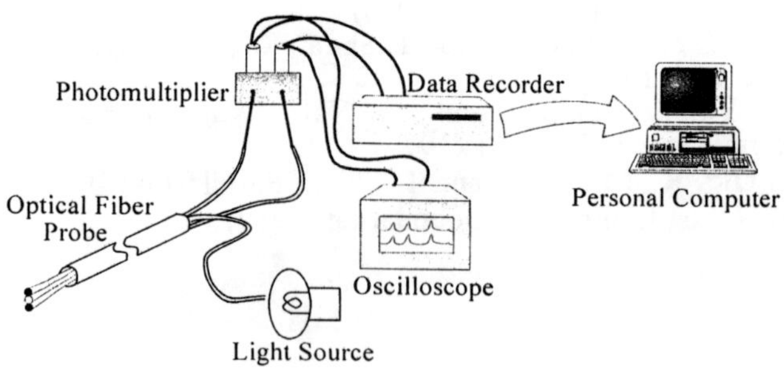

Fig.3 Schematic of data processing system.

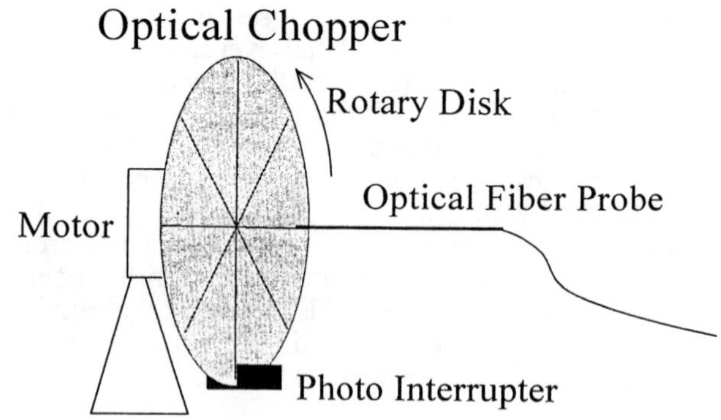

Fig.4 Schematic of rotary disk calibration system.

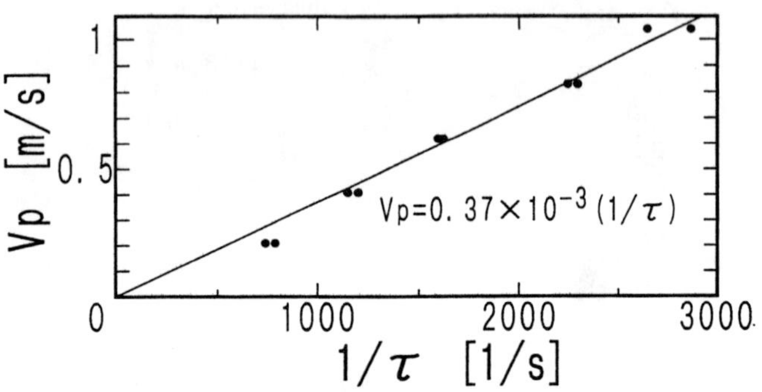

Fig.5 Correlation of v_p with $1/\tau$.

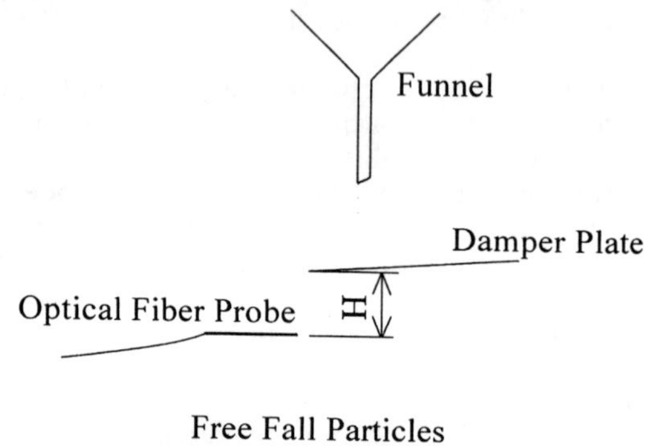

Fig.6 Schematic of free-fall particle calibration system.

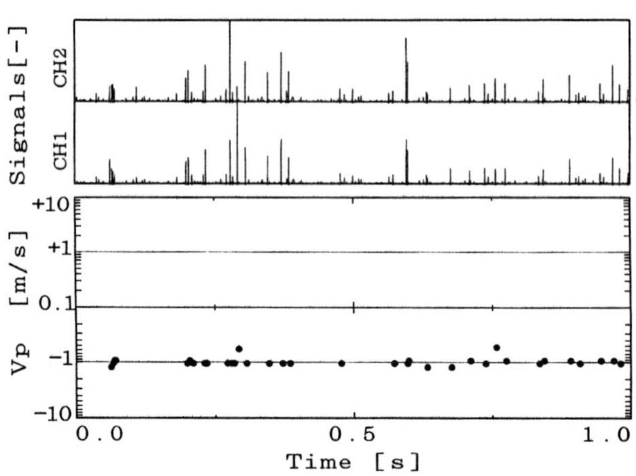

Fig.7 Probe signals and particle velocities for free-fall particles.

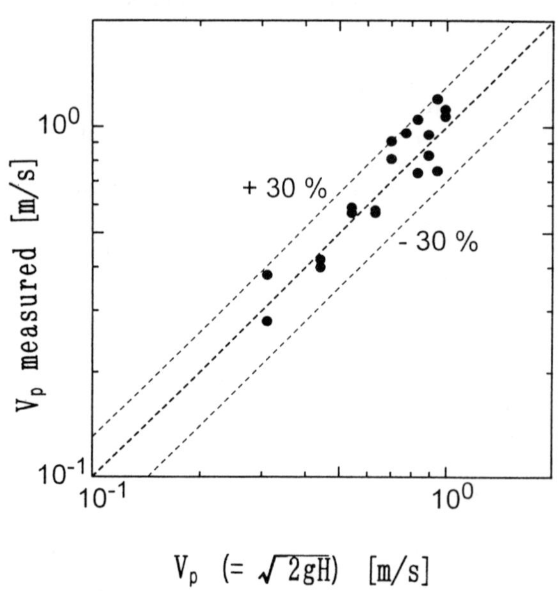

Fig.8 Comparison between v_p and calculated velocity.

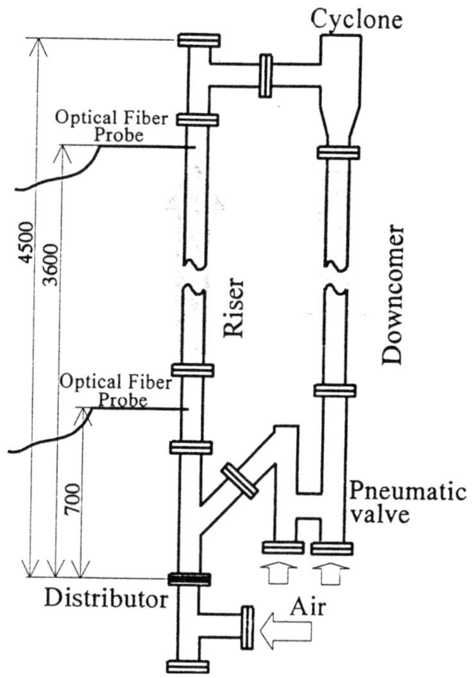

Fig.9 Schematic of circulating fluidized bed (CFB).

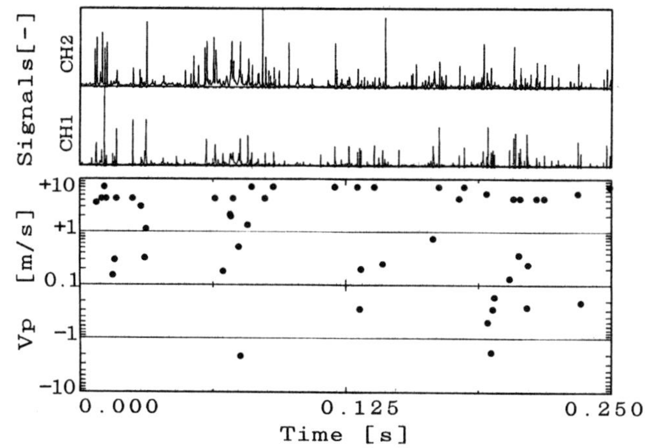

Fig.10 Probe signals and particle velocities in the upper central region of CFB.

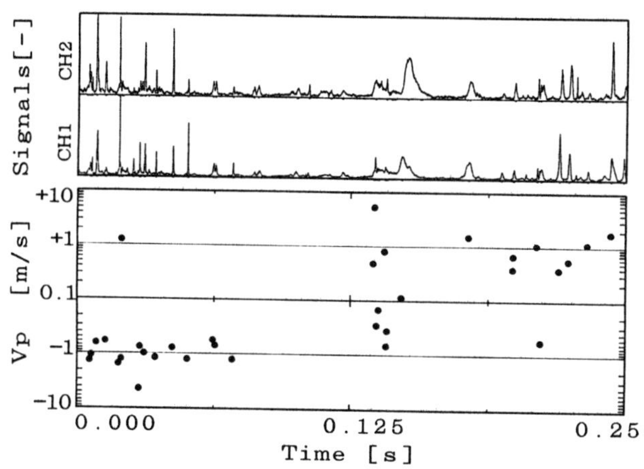

Fig.11 Probe signals and particle velocities in the upper wall region of CFB.

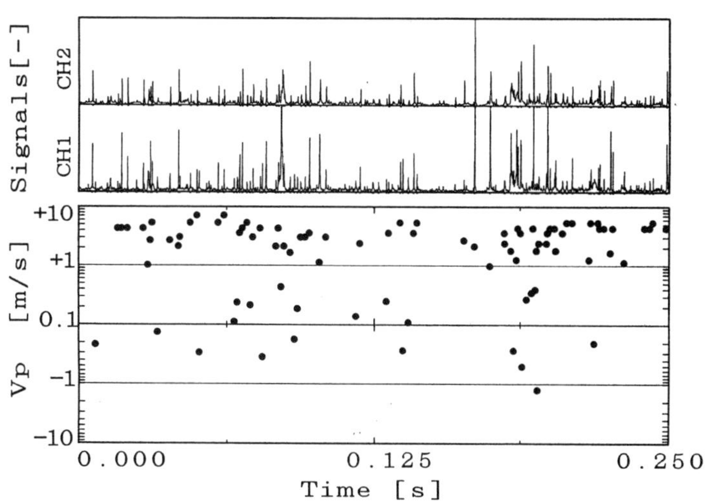

Fig.12 Probe signals and particle velocities in the lower central region of CFB.

On the Fluidization Behavior of Some Industrial Catalysts at High Temperature: The Effect of Interparticle Forces

Paola Lettieri, John G.Yates
Department of Chemical Engineering, University College London, UK
David Newton
BP Amoco Chemicals, Sunbury on Thames, UK

This paper reports some preliminary results on the effect of temperature on the fluidization of three fresh FCC catalysts and an equilibrium (E-cat) FCC tested at simulated process conditions, i.e. ambient pressure and at temperatures up to 6500C. Differences are highlighted in the fluidization of these powders between ambient data and elevated temperatures. The deaeration rate of the E-cat FCC increased with increasing temperature and the material became less aeratable due to the dominant role of the interparticle forces (IPF) over the hydrodynamic forces. Experimental results are compared with theory. Where changes in fluidization at high temperature were observed, the factors responsible are investigated. Results obtained from a thermo-mechanical analysis are also discussed. Some preliminary results on the fluidization quality of a Group A silica material doped with potassium acetate are also reported. These results demonstrate how temperature can increase the effect of IPF, causing a Group A material to behave in a similar manner to a Group C material.

Fluidization is of prime importance in Fluid Catalytic Cracking units as several parts of the process are performed in the fluidized state. The flow of catalyst in an FCC unit can be unstable if the particles become defluidized in the regenerator unit and standpipes. Changes in the physical properties of the catalyst are often responsible for the fluctuations and instabilities in the catalyst flow. The prediction of catalyst flow behaviour at unit operating conditions is therefore of major importance.

Fluidization quality has been related to particle and fluid properties and the effect of temperature on fluidization has been extrapolated from changes in the properties of the fluidizing gas, such as density and viscosity. Geldart et al. [1,2] established correlations on the basis of low temperature tests to predict powder behaviour and these have also been used for predictions at high temperatures. The role of interparticle forces on fluidization is still not clearly understood and is the subject of much academic and industrial research. Foscolo and Gibilaro [3] developed a stability criterion based on the assumption that hydrodynamic forces are the controlling factor on the fluidization quality. On the other hand, Massimilla and Donsi' [4] and Mutsers and Rietema [5] were amongst the first to suggest that Van der Waals Forces can influence the stability of the bubble free regime. Rietema and Piepers [6] and Poletto et al. [7] investigated the influence of operating pressure on the role of the interparticle forces and related the results to the stability of fluidized systems.

This work reports observations of the effect of elevated temperature on interparticle forces while fluidizing industrial catalysts. Major changes can occur during fluidization of some materials as IPF become greater with increasing temperature. A qualitative explanation of the phenomena observed is developed.

EXPERIMENTAL

A 140mm diameter x 1000mm tall Inconel vessel was used for testing the powders at ambient pressure up to 650°C. The vessel was fitted with a very fine stainless steel distributor plate (pore size of 20μm), to ensure a good and even fluidization in the bed, and with a disengaging section to avoid losing any elutriated fine material. The

[1]Department of Chemical Engineering, University College, London, UK.
[2]BP Amoco Chemicals, Sunbury on Thames, UK.
Paola Lettieri is now with BP Amoco Chemicals.

bottom section (or windbox) was packed with 10mm ceramic balls to improve heat transfer. Two solenoid valves were used during the bed collapse experiments, one to cut the gas supply to the bed, the other one to vent the air trapped in the windbox section. Fluidizing gas, nitrogen, was measured with rotameters before being preheated while passing through the windbox section, which was wrapped with an 800 watt heating tape. High temperatures were achieved by means of flexible ceramic mat heaters wrapped around the vessel, and capable of providing an operating temperature up to 1250°C. A system of thermocouples allowed the temperature in the fluidized bed to be monitored. A differential pressure transducer was used to measure the pressure profile as a function of the gas fluidizing velocity to determine the experimental minimum fluidization velocity, U_{mf}.

The minimum bubbling velocity was measured by visual observation of the bed surface, using X-rays, and noting the velocity at which the first bubbles began to appear as fluidizing gas velocity was increased. Recording the bed height while increasing the gas flow rate allowed the bed voidage to be measured.

The X-ray technique provides a moving image of the internal flow patterns of the fluids and solids inside the vessel, captured on a video tape at 25 frames per second. A pulsed high-energy beam is passed through the test vessel where attenuation depends on the quantity of fluids and solids within the vessel. The beam is synchronised with a video camera, thus allowing rapid changes in the internal flow patterns to be continuously detected and recorded.

The properties of the powders are summarised in Table 1.

Table 1 Powder	ρ_p kg/m^3	Geldart Group	d_p µm	F_{45} %
FCC 1 (Fresh)	1400	A	71	5
FCC 2 (Fresh)	1400	A	57	18
FCC 3 (Fresh)	1400	A	49	24
FCC 4 (E-cat)	1440	A	54	21
Doped Silica	1200	A	54	23

The mean particle diameter, particle size distribution and fines content of the powders tested were determined using a laser-light scattering technique (Malvern Instruments). Particle density was determined by using mercury intrusion analysis.

PRELIMINARY RESULTS ON FOUR FCC CATALYSTS

Dense phase collapse rate, U_{de}

Bed collapse profiles were obtained at increasing temperatures, the catalysts' response to ambient and high temperatures being compared with standard correlations. The bed collapse test was performed from ambient temperature up to 650°C, in steps of 100°C. At each temperature the nitrogen flow rate was set to give a constant gas velocity of about 1.5 cm/s (this velocity ensured a good mixing of the powder). Dense phase collapse rates (U_{de}) were obtained from the bed collapse profiles. U_{de} measures the rate of bed collapse in the absence of fluidizing gas. The bed collapse test relies on the premise that a fast rate of bed collapse is indicative of potential problems in an industrial unit. Experimental U_{de} values were compared with predictions obtained from Equation (1) developed by Abrahamsen and Geldart [1]:

$$U_{de} = \frac{0.314\,\rho_f^{0.023}(\rho_s - \rho_f)^{0.271}\,d_p^{1.232}\exp(0.508\,F_{45})}{\rho_f^{0.5}\,h^{0.244}} \quad (1)$$

Equation (1) predicts that U_{de} should decrease with increasing temperature, i.e. the dense phase should collapse at a slower rate as temperature rises. Experimentally all the fresh FCC materials behaved as predicted, with their deaeration rates significantly decreasing, see Figure 1: a 25% decrease occurred for FCC 3 between ambient temperature and 650°C, FCC 1 and FCC 2 both showed a 50% decrease.

The decrease in the dense phase collapse rate means that the capability of the catalysts to withstand a sudden change in the unit increased with increasing temperature.

On the other hand, FCC 4 became less aeratable with increasing temperature.

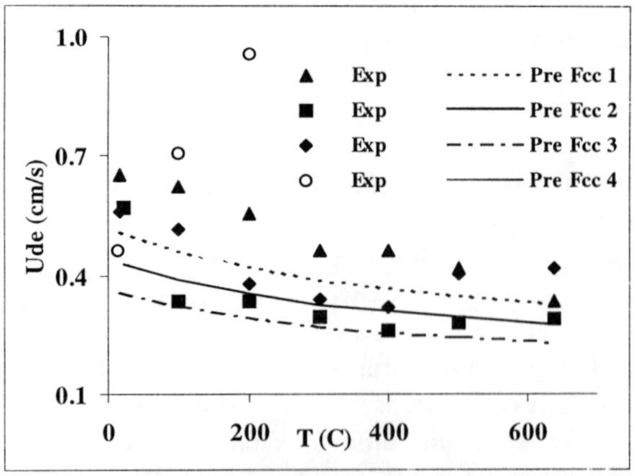

Figure 1 U_{de} *vs* T, comparison of experimental with Equation (1)

FCC 4 is an Equilibrium FCC, i.e. a catalyst which has been employed in several cracking and regenerating cycles in an FCC unit. A complex chemical structure characterizes E-cat FCC catalysts. A typical E-cat FCC sample contains process residues such as oil, coke, and deposit of metals such as Ion, Nickel and Vanadium. Large amounts of low boiling compounds and higher molecular weight compounds, from C_4 to C_{20}, including some aromatic compounds, also characterise the chemical mixture of an E-cat FCC.

The fluidization behaviour of FCC 4 was studied without performing pre-treatments on the E-cat powder prior to experiments. Thus, although the physical properties classify the E-cat FCC as a Group A powder, it may not show the flow behaviour of a typical Group A material due to its complex chemical structure. It is also worth emphasising that the flow behaviour of an E-cat FCC greatly depends on the efficiency of the stripper unit to remove the process residues.

The dense phase collapse rate, U_{de}, of FCC 4 increased sharply up to 200°C as shown in Figure 1. Fluidization of this catalyst could not be achieved at temperatures above 200°C, irreversible defluidization occurring at this temperature. This catalyst thus showed major differences in fluidization behaviour, compared to fresh FCC catalysts, and a possible explanation for this behaviour is as follows. Heavy hydrocarbons, which are solid at ambient conditions, may have fused with increasing temperature causing liquid to be expelled from within the pores of the catalyst and the particles to agglomerate and defluidize. The catalyst could not be re-fluidized after the system had been cooled back to ambient temperature. The pressure drop across the bed remained at zero for all flow rates, the gas channelling through the more or less solid mass formed when the liquid coating responsible for the defluidization solidified as the bed temperature decreased. It was only possible to break up this solid material by using a strong solvent followed by sand blasting.

To investigate the factors responsible for the observed changes in fluidization, a thermo-mechanical analysis was carried out on all the catalysts using a dilatometer in which changes in the catalyst dimension with increasing temperature were measured. Each catalyst was heated to 1000°C at a rate of 10°C per minute, and was subjected to a static force of 0.2 Newton. The results reported by the thermo-mechanical analysis were compared with fluidization findings. Thermograms obtained for FCC 3 and for FCC 4 (the E-cat FCC) are shown below, see Figure 2a and Figure 2b respectively.

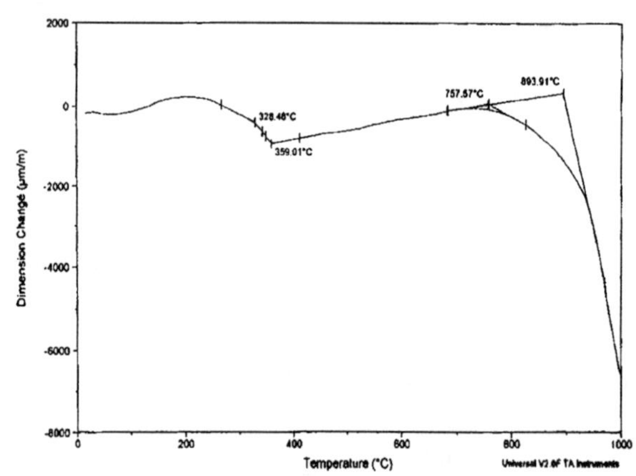

Figure 2a TMA - FCC 3 (Fresh FCC)

The FCC 3 thermogram showed an expansion of the material up to approximately 200°C followed by a fairly rapid shrinkage up to 360°C. Above this temperature the material showed linear expansion up to 750°C. Only above this temperature the material began to shrink again, possibly due to sintering (see Figure 2a).

No major changes occurred during fluidization of FCC 3, only a slight increase of the dense phase collapse rate was observed at temperatures above 300°C. Different thermograms were obtained for the fresh FCC 1 and FCC 2 catalysts compared to FCC 3. These materials showed an expansion up to approximately 200°C followed by a slow shrinkage up to about 500°C. Above this temperature the materials showed a faster rate of shrinkage up to 900°C. Only small differences were observed in the dimensional change at all temperatures between the thermograms of these two catalysts. Similar fluidization behaviour was also observed for both catalysts with increasing temperature, no significant changes occurred during fluidization of FCC 1 and FCC 2.

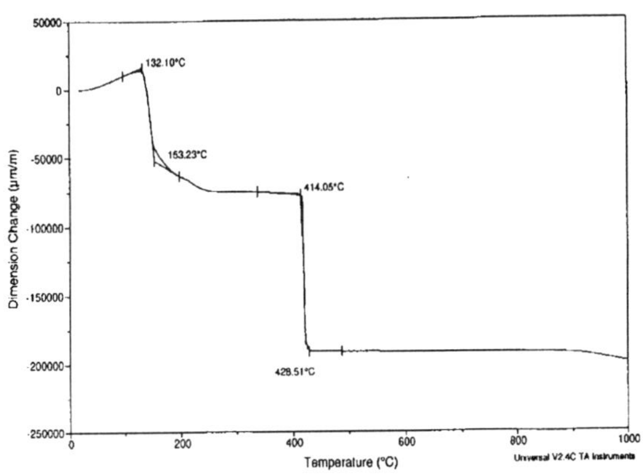

Figure 2b TMA - FCC 4 (E-cat FCC)

The thermogram for FCC 4, see Figure 2b, showed an initial expansion up to 134°C after which a sharp decrease in size occurred up to about 200°C. A drastic change in fluidization, within this range of temperatures, took place as shown in Figure 1 and discussed earlier. The thermogram also shows a second very sharp shrinkage between 414°C and 429°C, after which the size remained constant. As reported no experimental observation in the fluidized bed could be made at temperatures above 200°C.

Voidage at minimum bubbling, ε_{mb}

The effect of temperature on the minimum bubbling voidage was also investigated. Results obtained for the fresh FCC catalysts were compared with the predictions of the Foscolo and Gibilaro [3] criterion. Using Wallis' theory of wave propagation, they derived the condition for the voidage at the bubbling point as:

$$\left[\frac{(\rho_p - \rho_g)}{\rho_p}\right]^{0.5} \frac{\sqrt{gd_p}}{u_t} = 0.56n\varepsilon_{mb}^{n-1}(1-\varepsilon_{mb})^{0.5} \quad (2)$$

Their criterion was derived from theory based solely on hydrodynamic considerations and predicts that ε_{mb} should increase with increasing temperature, largely due to increasing gas viscosity. Our results show little change with temperature, in contrast with those obtained by Rapagna' et al [8] who also verified the validity of the criterion at high temperature.

Figure 3 shows a comparison of our data with the Foscolo and Gibilaro prediction. In equation (2) values of u_t were obtained using the Stokes' law, and the index "n" was calculated from correlations given by Richardson and Zaki [9]. It is worth noting that the smaller the particle diameter the bigger the over-prediction. Similar findings were also reported by Xie [10].

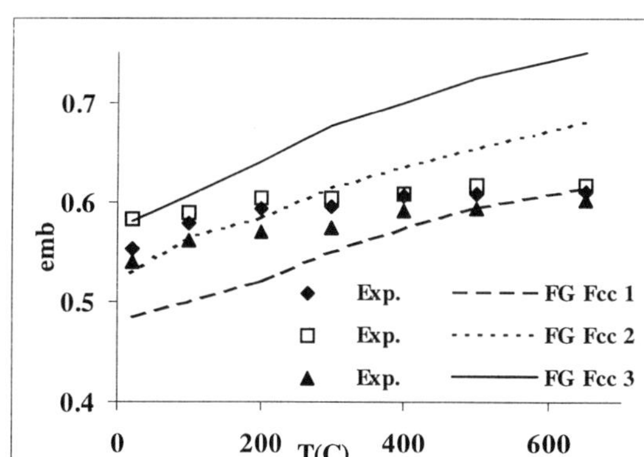

Figure3 ε_{mb} versus T, comparison of experimental with FG prediction

PRELIMINARY RESULTS ON A DOPED SILICA MATERIAL

From our results it appeared clear that the IPFs can play a major role on the fluidization of powders at high temperatures. Towards quantification of the effect of the IPFs we have tried to change the characteristics of the catalyst surface in a controlled manner. A batch of approximately 5 kg of a silica material was doped with some potassium acetate up to 10%wt. The potassium acetate was dissolved in distilled water to equal the total pore volume of the powder. The aim of the doping was to fill the pores of the catalyst. 1 kg samples were mixed thoroughly with the prepared solution. Each sample was dried prior to fluidization. An XRF (X-ray fluorescence spectroscopy) analysis was used to quantify the amount of potassium acetate present on the doped material. Details of this material are reported in Table 1.

The ratio between the measured and calculated pressure drop across the bed of material is plotted against gas velocity in Figure 4, as temperature is increased.

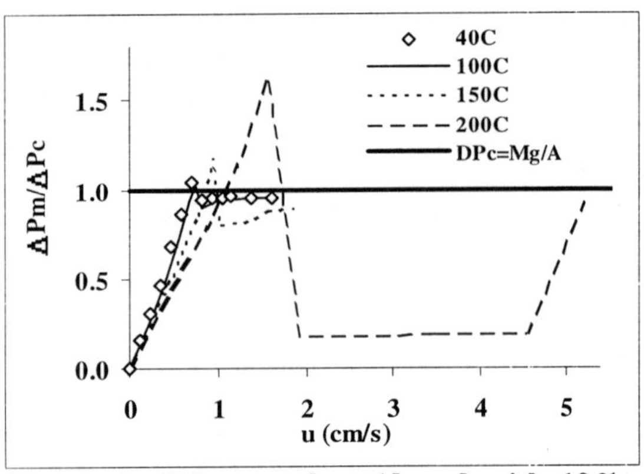

Figure 4 Silica catalyst (doped with 10%wt KOAc), $\Delta P_m/\Delta P_c$ *vs* Bed velocity

Fluidization behaviour of this material significantly changed when the bed temperature was about 200°C. A rapid decrease of the pressure drop was noted and channels were seen on the X-ray images. This material exhibited a typical cohesive Group C behaviour. Fluidization at 200°C could be achieved only when operating the system at a very high velocity, thus causing the channels and rat holes to be broken down. When the bed was cooled back to 150°C homogenous fluidization was obtained again at low velocities. It is possible that on increasing temperature the potassium acetate contained within the catalyst pores became mobile and migrated to the surface of the particles due to decreases in the surface tension and viscosity. The presence of potassium acetate on the surface of the catalyst caused the material to become sticky and to channel. On the other hand as the bed temperature decreased, re-absorption of the potassium acetate into the catalyst pores may have occurred, thus allowing homogeneous fluidization of the material to be re-established. The doped material thus showed similar behaviour to the E-cat powder at high temperatures but different behaviour on cooling when no solidification of the bed mass was observed.

CONCLUSIONS

Four FCC catalysts were characterised at elevated temperatures up to 650°C. Deaeration rates of the catalysts are reported and compared with the correlation of Abrahamsen and Geldart. Prediction was satisfactory for the fresh FCC catalysts. For the E-cat FCC however, IPFs increased with increasing temperature, the bed of particles becoming defluidized at 200°C. A possible explanation for this behaviour in terms of the nature of the forces acting between the particles is given. An attempt to relate fluidization findings at high temperature with results obtained from thermo-mechanical analysis is also described. Fluidization of a doped material at high temperature demonstrated the role of IPFs in making a Group A powder behave in a similar manner to a Group C material, and a further possible explanation for this is given.

ACKNOWLEDGEMENTS

The authors wish to acknowledge financial support provided for this work by BP Chemicals, Sunbury, and the EPSRC for a CASE studentship. Technical support at UCL and at BP Chemicals for this work is acknowledged. We are also very grateful to Dr. Michele Fiorentino for invaluable comments and discussions.

NOTATION

d_p	mean particle diameter	m
F_{45}	size fractions below 45μm	
h	settled bed height	m
n	Richardson and Zaki index	
U_{de}	dense phase collapse rate	$m\ s^{-1}$
u	superficial gas velocity	$m\ s^{-1}$
u_t	terminal fall velocity	$m\ s^{-1}$
ΔP_c	calculated bed pressure drop	$N\ m^{-2}$
ΔP_m	measured bed pressure drop	$N\ m^{-2}$

Greek Letters

ε_{mb}	minimum bubbling voidage	
ρ_g	gas and particle density	$kg\ m^{-3}$
ρ_p	particle density	$kg\ m^{-3}$
μ_g	gas viscosity	$N\ m^{-2}\ s$

LITERATURE CITED

1. **Abrahamsen, A.R., and Geldart, D.,** "Behaviour of gas-fluidized beds of fine powders Part II. Voidage of the dense phase in bubbling beds," *Powder Technology,* **26**, pp. 47-55 (1980).
2. **Geldart, D., and Wong, A.C.Y.,** "Fluidization of powders showing degrees of cohesiveness-II. Experiments on rates of de-aeration," *Chem. Eng. Sci,* **40,** pp. 653-661 (1985).
3. **Foscolo, P. U., and Gibilaro, L. G.,** "A fully predictive criterion for the transition between particulate and aggregate fluidization," *Chem. Eng. Sci.,* **39** (12), pp. 1667-1675 (1984).
4. **Massimilla, L., and Donsi', G.,** "Cohesive Forces Between Particles of Fluid-Bed Catalysts," *Powder Tech.,* **15**, pp. 253-260 (1976).
5. **Mutsers, S.M.P., and Rietema, K.,** "The effect of interparticle forces on the expansion of a homogeneous gas-fluidized bed," *Powder Tech,* **18**, pp. 239-248 (1977).
6. **Rietema, K., and Piepers, H.W.,** "Effect of interparticle forces on the stability of gas-fluidized beds-I. Experimental evidence." *Chem.Eng. Sci.,* **45**, pp. 1627-1639 (1990).
7. **Poletto, M., et al.,** "Fluidization of solids with CO_2 at pressures and temperatures", *Chem.Eng. Sci.,* **48**, pp. 617-621 (1993).
8. **Rapagna',S, et al.,** "The influence of temperature on the quality of gas fluidization", *Int. J. Multiphase Flow* **20** (2), pp. 305 (1994).
9. **Richardson, J.F., and Zaki W.N.,** "Sedimentation and Fluidization: Part I," *Trans. Instn. Chem. Engrs.* **32**, pp. 35-53 (1954).
10. **Xie, H.Y.,** "Fluidization of fine particles," *Ph.D. Dissertation*, University of Bradford, UK (1993).

Fluidized Bed Combustion - Areas of Development

Bo Leckner
Department of Energy Conversion, Chalmers University of Technology, 412 96, Goteborg, Sweden

The aim of this paper is to give a short survey of some selected areas of development of fluidized bed combustion. A few items are selected: desulfurization in pressurized applications, the role of volatiles and fuel size. It is concluded that there has been a dramatic increase of knowledge on fluidized bed combustion during the last decade, but there is still a need for a detailed description of the processes taking place in a commercial-size combustor as a basis for optimization of emission and combustion behavior.

Fluidized bed is a highly adaptable device and even in the limited field of combustion there are many different opportunities of design, dependent on energy supply and demand and on legislation and taxation policies in various countries. Therefore, subject to different requirements, fluidized bed combustors vary in type and application, as is also reflected in the number of installations in different countries, Figure 1[1] and in the items treated below. Owing to space limitations only a few items are selected for discussion: desulfurization in pressurized applications, scale-up, the role of volatiles and fuel size, and a number of important subjects are left out: ash behavior, emissions, waste fuels, etc, etc.

DESULFURIZATION – SOME RECENT WORK

The use of limestone for in-furnace desulfurization is one of the advantages of fluidized bed combustion (FBC). The goal is to reduce limestone consumption for a given degree of desulfurization and, in the combined cycle application which will be treated here, to avoid calcium sulfide (CaS) in the effluent ashes. During the last years the desulfurization properties of limestones for use in pressurized FBC (PFBC) have been intensely studied. The behavior of uncalcined limestone in a PFBC as well as sulfidation of limestone in a gasifier followed by sulfation and combustion of the remaining char have received attention. The PFBC is now well known, but there are several combinations of gasifiers and combustors, not yet introduced commercially.

Three combinations have been frequently mentioned:

1. Mild gasification of coal in a fluidized bed gasifier and combustion of the product gases in a gas-turbine combustion chamber together with the oxygen-containing gas from the PFBC. The char residue from the gasifier is burned in the PFBC which also serves as a heat source for a steam cycle. The gas is cleaned in a high-temperature ceramic filter [2,3].
2. Similar to 1 but the char from the gasifier is burned in an independent atmospheric circulating FBC (CFBC) connected to a steam cycle [4].
3. Similar to 1 but the flue gas from the PFBC is

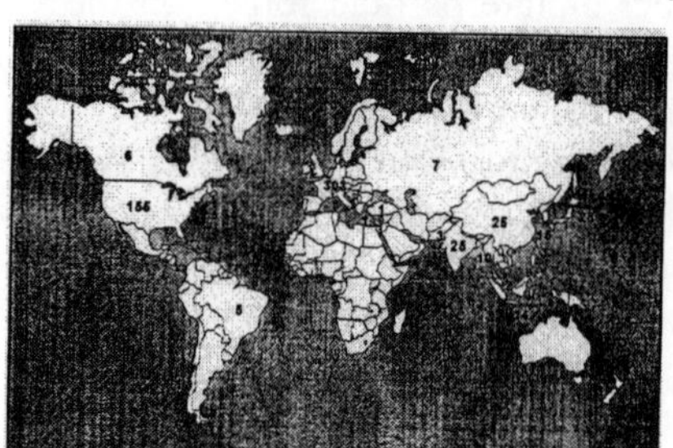

Figure 1. Number of FBC installations [1].

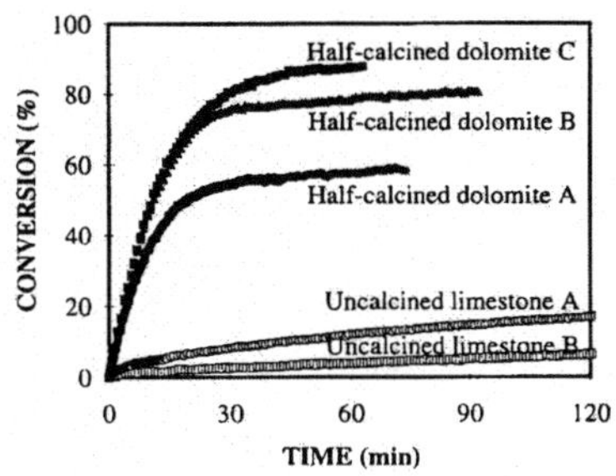

Figure 2. Conversion to CaS under gasification conditions. From [6].

introduced into the gasifier where most of the fuel conversion takes place. This gas leaves the gasifier together with the produced gas and is cooled to about 500°C before it is filtered in order to avoid using a not yet reliable high-temperature filter. Like in the other cases the gas is burned in the gas-turbine combustion chamber to attain a temperature level suitable for the gas turbine [5].

Desulfurization with limestone or dolomite is one of the critical processes in this application. The desulfurization process is limited by thermodynamic restrictions and by product layers formed by reactions. In a gasifier a few hundred ppm of the H_2S produced will not be reacted because of thermodynamics, but the rest reacts with calcined CaO or half-calcined dolomite, Figure 2. The principal overall reactions are

$$CaCO_3 \rightarrow CaO + CO_2 \text{ (calcination)} \quad (1)$$

$$CaCO_3 \cdot MgCO_3 \rightarrow CaCO_3 + MgO + CO_2 \text{ (half-calcination)} \quad (2)$$

$$CaO + H_2S \rightarrow CaS + H_2O \text{ (sulfidation)} \quad (3)$$

$$CaCO_3 + H_2S \rightarrow CaS + H_2O + CO_2 \text{ (direct sulfidation)} \quad (4)$$

The last reaction is only useful with half-calcined dolomite, so in general calcination is desired, and this means gasifier temperatures above 900 to 950°C depending on the CO_2 partial pressure.

In the combustor sulfation will take place without release of SO_2 if the temperature is < 900°C,

$$CaS + 2O_2 \rightarrow CaSO_4 \quad (5)$$

At temperatures >900°C and low O_2 concentrations SO_2 can be released from the $CaSO_4$ formed,

$$CaS + 3CaSO_4 \rightarrow 4CaO + 4SO_2 \quad (6)$$

and perhaps also at still higher temperatures

$$CaS + 3/2\ O_2 \rightarrow CaO + SO_2 \quad (7)$$

The high molar volume of $CaSO_4$ leads to formation of a product layer that prevents penetration of further oxygen to react with the core of CaS in a particle. Also, the exothermal reaction (5) may initially rise the temperature of the particle to favor reaction (6) [7], but eventually a core of CaS remains unreacted, surrounded by a shell of $CaSO_4$ and perhaps also CaO (the latter reacts to some extent with the SO_2 released). Abbasian et al.[8] summarizes their results regarding sulfation claiming that an optimum conversion from sulfide to sulfate is obtained in the temperature range of 815 to 900°C. Furthermore, the conversion is better for partially sulfided particles, i.e. the calcium to sulfur molar ratio(Ca/S) should be high in the gasifier. He also claims, in agreement with subsequent investigations, that the conversion to $CaSO_4$ is higher for dolomites having a high fraction of $MgCO_3$.

The laboratory investigations indicate that it may be difficult to convert all CaS in the combustor at the same time as a low amount of wastes is produced from the process (low Ca/S and no MgO).

Only a few pilot scale tests are reported in the literature. Foster Wheeler uses a gasifier at 925°C and a PFBC at 870°C. They attain a sulfur capture of 98.5% (probably at Ca/S=2) using a "proprietary sulfur capturing techniques" [9], so it is not clear what was actually done. No information concerning the CaS content in the ashes was given. However, in a previous work [10] a conversion of sulfide to sulfate of only 74 to 82 percent was attained. It was believed that the conversion would be higher in a larger unit. A comprehensive test on combustion of CaS-containing char from a gasifier has been carried out by Sage et al. [11]. Their atmospheric CFB combustor was operated at 3 m/s, 900 to 970°C and 6 to 37% excess air. Remarkably good (considering the high temperature) desulfurization efficiencies of above 96% were obtained for Ca/S of about 2 and the combustion efficiency was equally good, above 99%. The influence of air supply and temperature was small. However, the requirement of complete CaS conversion was difficult to fulfil. In line with the results of the laboratory investigations, dolomite proved to be more favorable than limestone; less than 1% of the sulfur remained in the form of sulfide in tests with dolomite, whereas 7% remained in tests with limestone, especially in the coarse bed material.

Optimization of limestone size and investigation of the capacity of different limestone types to reduce the problem of CaS inclusion is obviously desirable.

SCALE-UP

Are pilot plant results like those mentioned above valid for a large commersial plant? It is not possible to satisfy both chemical and fluid dynamic scaling criteria. Normally fluid dynamic scaling is sacrificed and chemical scaling is deemed to be satisfactory if the operating parameters are the same as in a large plant, and as stated in [11] "the gas residence time (is) typical of full scale units". These criteria usually result in tall but narrow test reactors.

Sulfur capture is sensitive to the combustion conditions in a furnace. Therefore this process can be taken as an indicator of the combustor conditions in a discussion about scale-up. Thermodynamics reveal that conversion of CaO and SO_2 , as well as that of CaS, goes towards $CaSO_4$ if sufficient oxygen is present. Under oxygen-deficient conditions, which could occur locally even at overall air excess, the reverse reactions yielding SO_2 may take place, especially in the presence of CO and H_2. According to the "oxygen depletion theory" [12] the well known temperature dependence of sulfur capture (giving a maximum at 850ºC) is caused by an increasing combustion of gases (especially volatiles) in the particle phase (where the lime is) at increasing temperatures. This combustion removes oxygen and impedes sulfur capture. A release of already captured sulfur could even occur. This phenomenon is assumed to take place in a bubbling bed, and it should be effective in the dense bottom bed of a CFB. In the case of char combustion without volatiles, such as in [11], this effect could be less important for two reasons: 1) There is less gaseous combustion in the case of char and the combustion becomes more evenly distributed in the bed. 2) The 0.15m diameter test plant operated at 3 m/s could have been in a turbulent regime of fluidization which results in better mixing between fuel and air than in a bubbling bottom bed of a commersial CFB. This could result in a sufficient difference between the test plant fired with char and a full scale plant burning a bituminous coal to explain the excellent sulfur capture at high temperature and relatively low excess-air ratios in the pilot plant [11] in contrast to larger plants [13] where temperature and air supply have a substantial influence. The presence of volatiles and the size of the plant explain these differences. In the large plant the release of volatiles and the distribution of fuel lead to inevitable irregularities in the local fuel-oxygen ratio. Such irregularities are minimized in a small cross-section reactor. This discussion is supported by [14] where desulfurization in a pilot plant was compared with results from corresponding fuels in commersial boilers. It was found that with low-volatile fuels desulfurization versus Ca/S was similar, whereas for high volatile fuels the sulphur capture in the pilot plant was almost twice as efficient as in the large boilers.

In general, not only limited to sulfur capture, the cross-sectional effects cause differences between small and large plants: the larger cross-section suffer from an imperfect mixing of fuel and air and further experience is needed to transfer results from pilot scale [15]. Returning to the question posed initially, the conclusion becomes: a large atmospheric CFB burning char can have almost as good a sulfur capture as a pilot plant if the air and fuel supplies are designed for good mixing of fuel and air, but this would be difficult to achieve with fuels containing substantial amounts of volatiles.

In the case of a pressurized combustor the problems discussed have less importance for sulfur capture, because the higher SO_2 partial pressure results in a thermodynamically more favorable situation.

VOLATILES

Volatile combustion causes oxygen deficient regions which affect not only desulfurisation but also char burn-up, deposition of ashes on surfaces and corrosion. The volatile content and the size of fuel particles are important for the mixing of fuel and air as well as for the allocation of fuel feed points and the method of fuel feed. In the worst case a rapid release of volatiles may have the effect illustrated on Figure 3. A fuel-rich gaseous stream created at the fuel feed point may remain relatively unaffected during its passage through the combustion chamber as illustrated by measurements [16]. Obviously, gas mixing in CFB is not good (as it was claimed to be in early publications). It is quite similar to dispersion

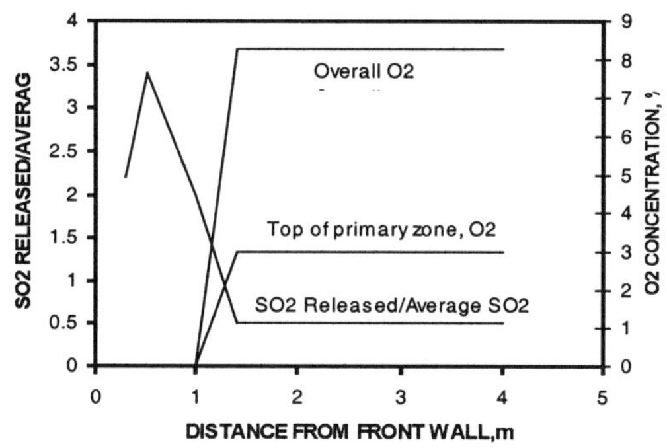

Figure 3. Release of SO_2 in the vicinity of a fuel feed point. From [14].

in turbulent one-phase flow, Figure 4 [17]. The problem is usually not noticed outside the.combustion chamber in the form of excessive emissions of unburned gases because of the beneficial effect of mixing in the cyclone, but it can explain the performance of a boiler with respect to sulfur capture efficiency, combustion efficiency etc.

For high volatile fuels (biomass and waste) burned in small boilers, the stationary (bubbling) fluidized bed is often preferred [18]. The effects of volatiles combustion are similar to what was indicated above: over-bed combustion takes place, partly because of insufficient mixing and combustion in the bed. In a non-circulating bed this can be better handled by mixing arrangements in the freeboard, since particles are not present. Because of

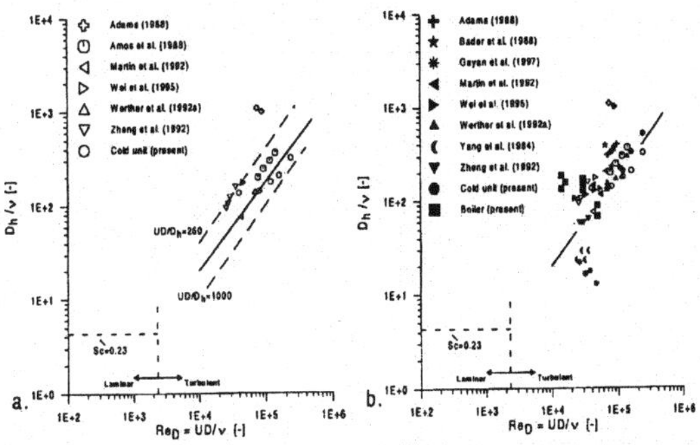

Figure 4. Dispersion of gas in CFB risers. Solid line literature data from single phase flow. Open symbols empty CFB risers. Black symbols risers with particles. From [17].

the similarity to grate firing, techniques have been borrowed from such furnaces to provide mixing: mixing zones and noses, as illustrated in Figure 5.

Conclusion: the mixing problem caused by volatiles has an importance for many processes in a CFB

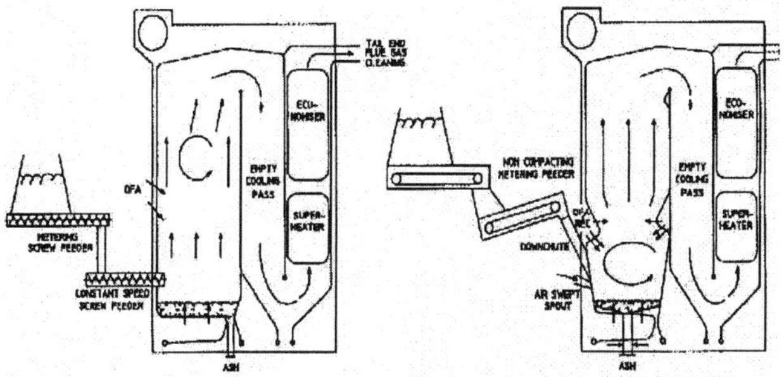

Figure 5. Development of FBC boilers for waste combustion. From Kvaerner [19].

burning coal, but it is clearly decisive for combustion of biofuels and high-volatile wastes.

PARTICLES

After fragmentation, attrition and combustion the materials supplied to the bed produce a certain particle size distribution in the bed. This size distribution results in a corresponding vertical suspension density distribution which affects heat transfer to the walls and determines bed temperature: if the size distribution becomes biased towards large particles, the particle density in the heat transfer region decreases, which reduces the heat transfer, and the bed temperature increases. This influence of bed material size has been clearly explained [20]: the small particle side of the spectrum is determined by the fractional efficiency of the gas-particle separator, whereas the coarse particle side of the spectrum is found in the bottom bed and has to be adjusted by bed drainage. Various types of bed material regeneration devices have been invented to remove the coarse particles and send back the fine fractions to the bed to avoid loss of bed material.

Despite this general knowledge, problems have occurred after changing from a fuel producing fine ashes to fuels containing stones [21]. The solution was to remove bottom bed material for regeneration by crushing and return the crushed material to the bed. This caused the ashes to leave the boiler almost entirely in the form of fly ash. However, stones could pile up on the bottom of the bed, and such piles are difficult to move: treating the fuel before it is fed into the boiler solves part of this problem [22]. Another problem was found in the 250 MW boiler in Gardanne, where the heat transfer to the boiler walls was insufficient. To control bed temperature, heat had to be extracted by the superheater surfaces in the external heat-exchangers, which resulted in an extreme need for steam cooling to reach the desired steam temperature. The remedy [23] was to lengthen the vortex-finder of the cyclones somewhat and to displace it into an eccentric position in the cyclone according to Muschelknautz [24]. These measures enhanced the cyclone efficiency on the small size side, which changed the vertical particle suspension density distribution and the heat transfer in the furnace increased, less heat had to be removed in the external heat exchangers and the operation of the boiler improved [23].

From these examples it is evident that the situation can be handled by design measures. But is there enough knowledge for a conscious optimization of particle size? The following items are of interest in this context:

1. Size of fuel and limestone. Large fuel particles require less communition work and improve the distribution of fuel across the cross-section of the furnace, but more efforts have to be spent on bed material regeneration. Optimization of bed material size requires improved knowledge on fragmentation and attrition and on dispersion of fuel under actual boiler conditions. The present abundant information is mostly limited to laboratory conditions.
2. Segregation effects. The movement of large particles in a bed of small particles applied to fuel particles in a boiler bed and to separation of large particles from small particles in a classifier. Again, work has been done in this field, but the results are difficult to apply at the desired operation conditions.
3. Gas-particle separators. The vortex finder in cyclones has been a constant source of preoccupation, either it is too long and erodes or it is too short and does not perform well. Also, many different shapes have been proposed for separators. The square separator of Foster Wheeler [25] is remarkable. If the efficiency is satisfactory in such a device, new possibilities arise to avoid the expensive cyclone. However tools are needed: numerical fluid dynamic calculations are suitable for this type of investigation. Single particle tracking gives some information, but the goal of development of computational techniques is to handle both particle-particle interactions and particle size distributions.

In conclusion, optimization of particle size (fuel, sorbent, bed) would relieve several secondary problems caused by local unmixedness in FBC boilers. Such an optimization needs adequate information on fuel and sorbent behavior inside the combustor and in the external particle separation devices.

CONCLUSION

During this decade there has been a dramatic increase of knowledge on FBC design as well as of detailed phenomena. However too little scientific attention has been directed to the processes taking place on the scale of a combustion chamber, and processes such as volatile combustion and fuel dispersion are not well known on the scale of a combustor. This in turn is important for evaluation of the emission and combustion behavior of the boiler.

LITERATURE CITED

1. "The Future of Fluidized Bed Combustion", VGB Report 212e, *VGB Technische Vereinigung der Grosskraftwerkbetreiber*, Essen, pp. 6-20 (1997).
2. **Robertson, A.,** *9th Int. Conf. on FBC,* J.P.Mustonen ed., ASME, New York, pp. 1013-1020 (1987).
3. **McClung, J.D.,Kastner, C.,Dellefield, R., and Sears S.,** *12th Int. Conf. on FBC,* L.N.Rubow and G.Commonwelth, eds., ASME, New York, pp.1047-1052 (1993).
4. **Daves, S.G., Mordecai, M., Brown, D., and Burnard, G.K.,** *13th Int. Conf. on FBC,* K.J.Heinschel, ed., ASME, New York, pp. 1275-1282 (1995).
5. **Mori, S.**, *Fluidization IX,* L.S.Fan and T.M.Knowlton, eds., Engineering Foundation, New York, pp. 45-52 (1998).
6. **Yrjas P., Iisa K., and Hupa M.,** *Fuel* **75**,89-95 (1996).
7. **Davies N.H., Laughlin K.M., and Hayhurst A.M.,** *25th Symp.(Int.) on Combustion,* The Combustion Institute, Pittsburgh, pp. 211-218 (1994).
8. **Abbasian, J., Rehmat, A., and Banerjee, D.D.,** *Ind.Eng.Chem.Res.* **30**,1990-1994 (1991).
9. **Robertson, A., and VanHook, J.,** *14th Int. Conf. on FBC,* F.D.S.Preto, ed., ASME, New York, pp. 87-97 (1997).
10. **Conn, R., Van Hook, J., and Robertson, A.,** *13th Int. Conf. on FBC,* K.J. Heinschel, ed., ASME, New York, pp. 75-68 (1995).
11. **Sage, P.W., Schofield, P.A., and Gayán, P.,** *14th Int. Conf. on FBC,* F.D.S.Preto, ed., ASME, New York, pp. 847-856 (1997).

12. **Dennis, J.S., and Hayhurst, A.N.,** *20th Symp.(Int.) on Combustion,* The Combustion Institute, Pittsburgh, pp. 1347-1355 (1985).
13. **Lyngfelt, A., and Leckner, B.,** *Fuel* 72,1553-1561 (1993).
14. **Alliston, M.G., and Wu, S.,** *Circulating Fluidized Bed Technology,* M.Kwauk and J.Li, eds, Science Press, Beijing, pp. 327-332 (1996).
15. **Knöbig, T., Werther, J., Åmand, L.-E., Leckner, B.,** *Fuel* 77,*1635-1642 (1998).*
16. **Couturier, M.,Doucette, B., Stevens, D.,Poolpol, S., and Razbin, V.,** *11th Int.Conf.on FBC,* E.J.Anthony, ed.,ASME, New York, pp. 107-114, (1991).
17. **Sternéus, J.,***Gas Mixing in Circulating Fluidized Bed Risers*, Licentiate Thesis, Dept. of Energy Conversion, Chalmers Technical University, ISBN 91-7197-667-9 (1998).
18. **McCann, D.,** *14th Int. Conf. on FBC,* F.D.S.Preto, ed., ASME, New York, pp. 29-37 (1997).
19. **Gustafsson, L., Lundberg, M., and Victorén, A.,** *The Future of Fluidized Bed Combustion,* VGB Report 212e, VGB Technische Vereinigung der Grosskraftwerkbetreiber, Essen, Presentation 3.2 (1997).
20. **Herbertz, H., Lienhard, H., Barner, H.E., and Hansen, P.L.,** *10th Int. Conf. on FBC,* A.M.Manaker, ed., ASME, New York, pp. 1-6 (1989).
21. **Grief, H. G., Höhner, S., and Skiba, H.,** "The Future of Fluidized Bed Combustion", VGB Report 212e, VGB Technische Vereinigung der Grosskraftwerkbetreiber, Essen, Presentation 2.3 (1997).
22. **Tham, J., Vilmec, L., and Scheffknecht, G.,** *VDI Berichte* 1314, pp. 21-36 (1997).
23. **Frydrychowski-Horvatin, J., and Vostan, P.,** *VDI Berichte* 1314, pp. 5-20 (1997).
24. **Muschelknautz, U., and Muschelknautz, E.,***Circulating Fluidized Bed Technology,* M.Kwauk and J.Li, eds., Science Press, Beijing, pp. 597-602 (1996).
25. **Hyppänen, T., Raino, A., Kauppinen, K.V.O., and Stone, J.E.,** *14th Int. Conf. on FBC,* F.D.S.Preto, ed., ASME, New York, pp.183-190 (1997).

Dry Particle Coating by a Newly Developed Rotating Fluidized Bed Coater

Satoru Watano*, Robert Pfeffer, Rajesh N. Dave and William Dunphy
*Department of Chemical Engineering, Osaka Prefecture University, Osaka 599-853, Japan
Particle Technology Center, New Jersey Institute of Technology, Newark, N.J. 07102

Synthesis of composite materials has been conducted using a dry particle coating method via a newly developed rotating fluidized bed coater (RFBC). In this study, dry coating were conducted via the RFBC using PMMA particles (200 [tin] as core and fine Al_2O_3 (0.7 [tin]) as guest particles under various operating conditions. Properties of the coated products such as yield, sticking ratio of Al_2O_3 and specific surface area were investigated. Dispersion of Al_2O_3 on the surface of PMMA particle was also evaluated using an energy dispersive X-ray spectrometer (EDX). The feasibility of using the RFBC to coat dry particles was evaluated by the properties.

INTRODUCTION

The synthesis of composite materials with tailored properties has attracted special interest in many industries including pharmaceuticals, food, cosmetics, ceramics, electronics and special chemicals. Composite materials can be accomplished by both wet and dry processes, however, dry processes have received much attention because they are ecological and require no binder nor solvents. Thus, study of the dry processes has been desired for the development of particulate coating technologies.

Typically, dry methods to synthesize composite materials utilize some sort of grinding systems. They coat core particles using fine particles by subjecting the mixture to large shearing and compressive stress or high impaction forces. Up to this time, Yokoyama et al. [1], Naito et al. [2] and Alonso et al. [3] have studied the dry particles coating method to synthesis several composite materials using a high shear mill called Mechnofusion. Honda et al. [4] have analyzed the dry particle coating process using a dry impaction blending method named Hybridization system. Miyanami et al. [5] have also developed a horizontal high shear mill called Theta-composer for the particle dry coating. In all these devices, the high shear stress or impaction force may determine the depth to which the guest particles are embedded into the surface of the host particles. The local temperature rise due to these strong forces may also play a role in determining the coating strength. However, the high shear stress or the impaction force may not be favorable for some of pharmaceutical and food powders that are very weak to the heat.

In this study, we have developed a novel dry particle coating system using a rotating fluidized bed (RFB). Performance of the system was confirmed using polymethylemethacrylate (PMMA) particles for the core and very fine Al_2O_3 particles for the guest. Effects of the several operating variables on the surface properties of the coated products were evaluated quantitatively. Mechanism of synthesizing composite materials via the RFBC was also discussed here.

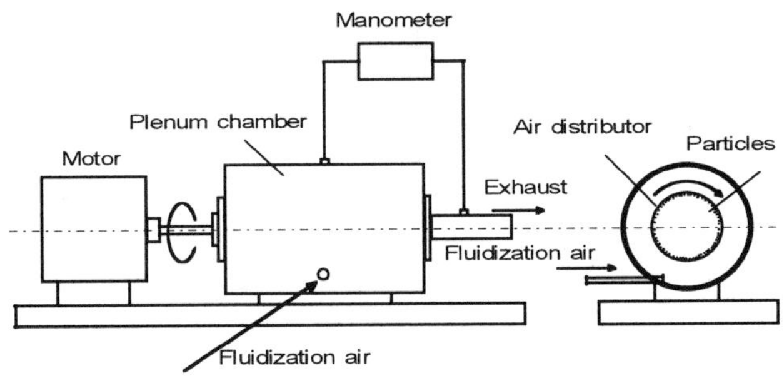

Figure 1 A schematic diagram of the experimental apparatus (RFBC)

EXPERIMENTAL

A schematic diagram of the experimental apparatus is shown in Figure 1. A rotating fluidized bed composes of a cylindrical plenum chamber and a slotted cylinder with fine screen (opening diameter is 43.2 μm) for the air distributor. The plenum chamber is of 101.6 mm inside diameter and 133.0 mm height, and the slotted cylinder is of 57.2 mm inside diameter and 88.9 mm height, respectively. The horizontal slotted cylinder with fine screen rotates around its axis of symmetry inside the plenum chamber. Core particles (PMMA: 200μm) and guest fine particles (Al_2O_3: 0.7 μm) are introduced into the cylinder and are forced to the wall due to the high centrifugal forces by the rotation. Air flows radially inward through the fine screen, and particles are balanced by the air flow and the centrifugal force, resulting in uniform particle fluidization. During the operation, the guest particles can be coated uniformly onto the surface of the core particles due to the furious fluidization.

After the coating operation, all of the coated products were taken out to measure their weight. Yield of the products was measured using the mass of particles before and after the coating. 10 g of the coated products were fed into a mesh basket (inside diameter is 15mm, height is 20mm, and the mesh opening is 75 μm) and sank into a water bath, subjecting to an ultrasonic vibration (1 min, 60 Hz). After removing loosely attached particles, the coated products were dissolved into an acetone. Since PMMA particles dissolve into acetone, Al_2O_3 particles could be separated from PMMA completely by using an ultra fine filter (pore size is 0.02 μm). Sticking ratio was then calculated using mass of Al_2O_3 on PMMA particles before and after the ultrasonic test.

Specific surface area was measured using a BET method. For each sample, 3 measurements were conducted to calculate the average value. Dispersion of Al_2O_3 on the surface of PMMA particles was analyzed by an EDX (Energy Dispersive X-ray Spectrometer). In this study, 30 g of PMMA and 1.5 g of Al_2O_3 (5 wt%) were used for the experiments.

RESULTS AND DISCUSSION

Figure 2 illustrates the pressure drop plotted against the airflow velocity at different rotational speeds. The pressure drop showed a rise to a peak at minimum fluidization velocity (u_{mf}) and then decreased slightly as air velocity continued to increase above u_{mf}. After that, it showed constant pressure drop regardless of the increase in air velocity. Due to the radial

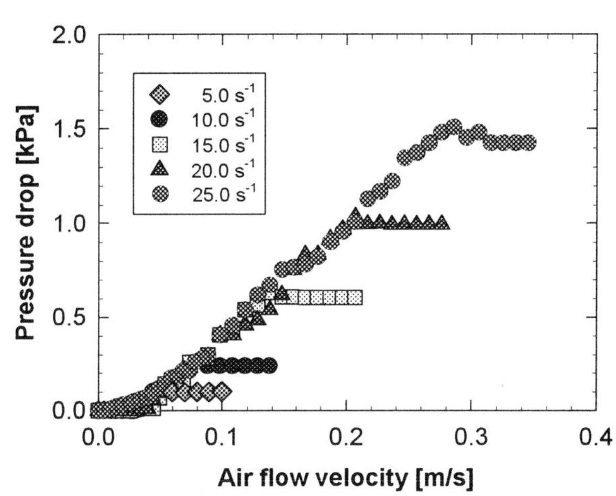

Figure 2 Pressure drop v.s. air flow velocity

acceleration, pressure drop increased with an increase in cylinder rotational speed.

Figure 3 shows the relationship between minimum fluidization velocity and cylinder rotational speed. Minimum fluidization velocity increased awfully with the rotational speed. Observed minimum fluidization velocity were in good agreement with the theoretical data calculated from the following equation proposed by Kao et al. [6]

$$\frac{u_{mf}\rho_f d_g}{\mu} = \left[\left(33.7\frac{C_2}{C_1}\right)^2 + 0.0408\frac{\rho_f(\rho_g - \rho_f)d_g^3\omega^2}{\mu^2}\frac{C_3}{C_1}\right]^{1/2} - 33.7\frac{C_2}{C_1}$$

where

$$C_1 = r_0^2(1/r_i - 1/r_0), \qquad C_2 = r_0\ln(r_0/r_i), \qquad C_3 = (r_0^2 - r_i^2)/2$$

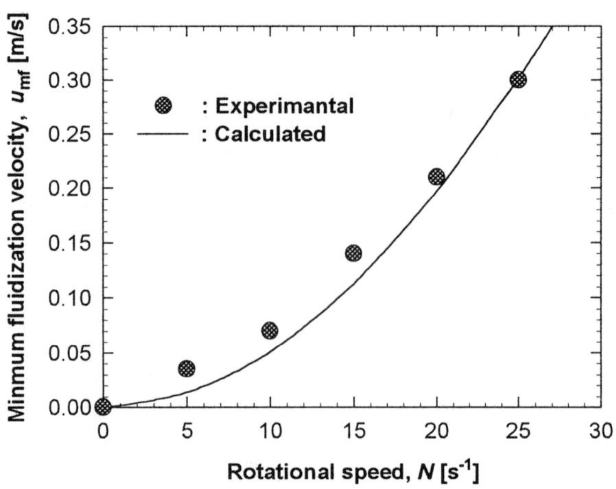

Figure 3 Minimum fluidization velocity v.s. rotational speed

Figure 4 indicates the plots of products' yield against the dimensionless air flow velocity at different rotational speeds. Before the bubbling condition ($u/u_{mf} < 2.0$), carry-over of PMMA particles were hardly observed during the operation. Almost 100 % of the PMMA particles remained in the bed during the operation. It could thus be assumed that the entrainmant of fine Al_2O_3 particles or formation of ordered mixture took place during this fluidization regime. Only the products at the lowest rotational speed (N=5.0s^{-1}) showed a little bit smaller yield than the other ones. It was supposed to be the disproportionate fluidization caused by the insufficient

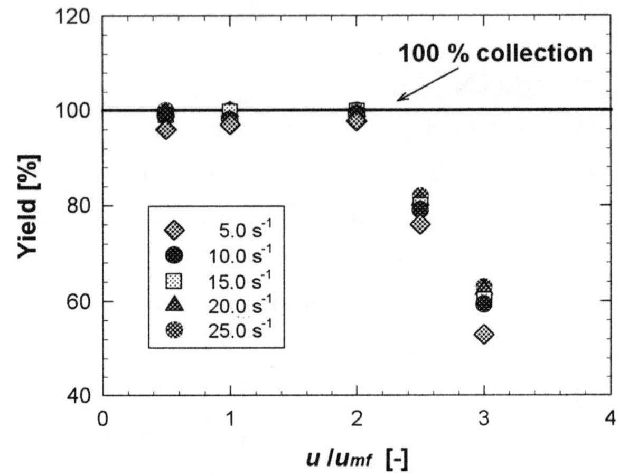

Figure 4 Yield of product v.s. u/u_{mf}

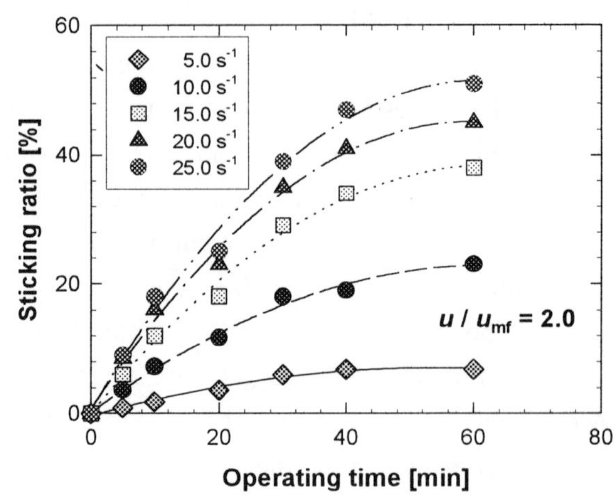

Figure 5 Sticking ratio v.s. operating time

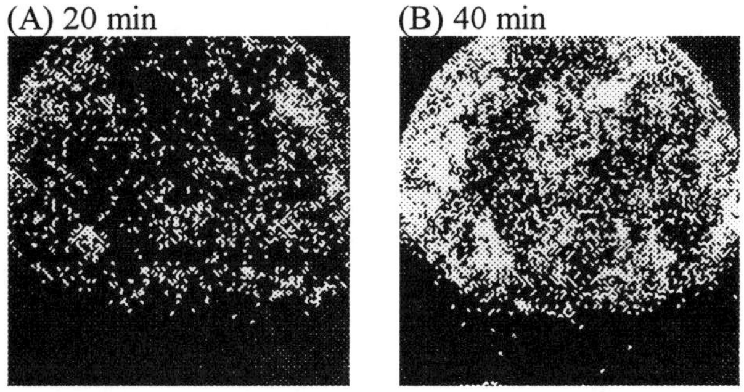

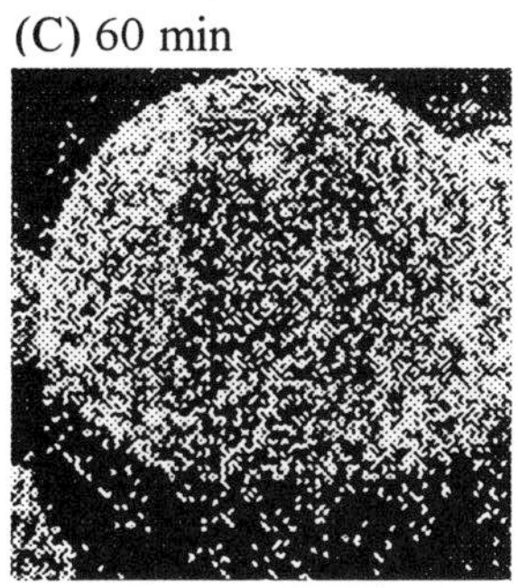

Figure 6
Al_2O_3 mapping by EDX
($\times$500)
($N = 20.0\ s^{-1}$, $u/u_{mf} = 2.0$)

cylinder rotation. However, in the region where the air velocity was greater than $2u_{mf}$, the yield of products decreased awfully with the air velocity. In the bubbling condition, a lot of particles were furiously thrown away by the bubble breakage, then carried away from the plenum chamber with the fluidization air. Therefore, the operating air velocity was determined to be $2u_{mf}$ because of the maximum products yield.

Figure 5 illustrates the sticking ratio plotted against the operating time. Most of the fine Al_2O_3 at the lowest rotational speed (N=5.0s^{-1}) were washed away by the ultrasonic vibration. This implied that most of the fine Al_2O_3 particles were stuck very loosely onto the PMMA particles just as the ordered mixture. As the rotational speed increased, however, the sticking ratio increased remarkably. Due to the furious agitation, particles collided each other intensely inside the powder bed. The impaction force at the collision became larger with the centrifugal force. Unlike the high shear mixer [1-3,5] or the impact blending method [4], there was no obvious heat generation. However, on the local contacting point between PMMA and Al_2O_3 particles, local heat was slightly generated, leading to the slight deformation of PMMA surface and Al_2O_3 particles. At the high centrifugal force, the local heat attached the Al_2O_3 particles more firmly. In fact, we have already confirmed that SEM picture of the PMMA cross section showed the Al_2O_3 particles were stretched and slightly embedded onto the surface of PMMA particles after the coating.

Figure 6 shows the EDX pictures of the composite surface. The fine Al_2O_3 particles were dispersed very well, and stuck firmly with deformation. According to the surface analysis by SEM, very few Al_2O_3 formed large agglomerate. Most of the particles were dispersed well as the primary size or a slightly agglomerated condition. These EDX pictures also made it clear that the dispersion condition became better with the elapse of operating time. It was found that the rotating fluidized bed worked very well not only for the dry coating but also for deagglomeration of sub-micron sized fine particles.

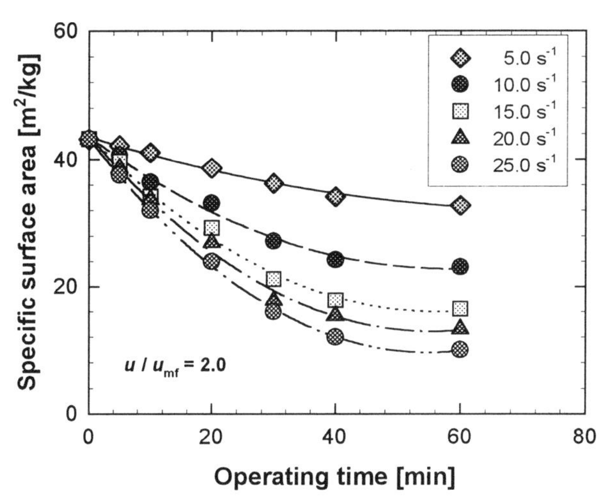

Figure 7 Specific surface area of coated products v.s. operating time

Figure 7 describes the specific surface area plotted against the operating time. In this figure, the data at t=0 showed specific surface area of the physical mixture of Al_2O_3 and PMMA particle. With the increase in rotational speed and operating time, the surface area decreased remarkably. This was because the Al_2O_3 particles dispersed and stretched well to form the smooth surface. These data also implied that the coating of fine particles were taken place in the rotating fluidized bed coating.

CONCLUSION

In this study, we have developed a novel dry coating system using a rotating fluidized bed. Performance of the system was evaluated by the properties of products obtained. It was found that the rotating fluidized bed could be used for particle dry coating.

ACKNOWLEDGMENT

The authors wish to thank Mr. Sanjiv Patel and Ms. Michelle Ramlakhan for their assistance in part of the experimental work.

NOTATION

d_g = granule diameter, m
N = rotational speed, s^{-1}
r_i = radius of inner surface of granule bed, m
r_0 = radius of rotating fluidized bed, m
u = air velocity, $m \cdot s^{-1}$
u_{mf} = average minimum fluidization velocity, $m \cdot s^{-1}$
μ = air viscosity, $kg \cdot m^{-1} \cdot s^{-1}$
ρ_f = air density, $kg \cdot m^{-3}$
ρ_g = granule density, $kg \cdot m^{-3}$
ω = angular velocity, $rad \cdot s^{-1}$

LITERATURE CITED

1. **Yokoyama, T., K.Uehara and M. Naito,**" The Angmill Mechanofusion System and its Applications," *Kona*, **5**, pp.59-68 (1987).

2. **Naito M., M. Yoshikawa, T. Tanaka and A. Kondo,** "Analysis of the Powder Composite Process by a Mechanical Method," *Kona*, **11**, pp. 229-234 (1993).

3. **Alonso M., M. Satoh and K. Miyanaimi,** "Mechanism of the Combined Coating-Mechanofusion Processing of Powders," *Powder Technol.*, **59**(1), pp.45-52 (Sept. 1989).

4. **Honda H., K. Ono, T. Ishizaka, T. Matsuno, T. Katano, and M. Koishi,** "Surface Modification of Powders by the High Speed Impact Tretement Method," *J. Soc. Powder Technol. Japan*, **24**(9), pp. 593-599 (Sept. 1987).

5. **Miyanami K., M. Satoh and T. Yoshida,** "A Novel Particles Composer and Its Performance Evaluation," Proceedings of First International Particle Technology Forum, AIChE Annual Meeting, Denver, CO, pp. 127-132 (Aug. 1994).

6. **Kao J., R. Pfeffer and G. I. Tardos,** "On Partial Fluidization in Rotating Fluidized Beds, " *AIChE Journal*, **33**(5), pp. 858-861 (May 1987).

Index

A

Analysis of some fluidized systems 1

B

Beads, solid glass 90
Bubble columns, studies of 30

C

Catalyst, FCC 42, 67,100
Centrifugal dedusters, efficiency curves for 13
Circulating fluidized beds (CFB) 42, 52, 61, 95
Cluster(s)
- catalyst 42
- formation 48

Coater, rotating fluidized bed 112
Combustion, fluidized bed 106
Computational Fluid Dynamics (CFD). 1,13,18
Computer automated radioactive particle tracking. .30
Cyclone separators 13
Cylindrical multiparticles, sedimentation of 36

D

Decomposition, ozone 52
Desulfurization in pressurized applications 106
Discrete element method (DEM) simulation 71
Down flow reactors, particle clustering in 42
Dry particle coating 112

E

Efficiency Curves, for centrifugal dedusters 13
Energy dispersive X-ray spectrometer (EDX) 112
ETH experiments 61

F

Fluctuation velocity 7
Fluid catalytic cracking (FCC). 18, 42, 67,100
Fluidized
- bed (s)
  - coater 112
  - combustion, areas of development. 106
  - DEM simulation of 71
  - gas, particle fluctuation velocity in 7
  - circulating, mass transfer in 52
  - olefin polymerization reactor 77
- systems, analysis of 1

Fluidization of industrial materials 100

G

Gamma ray computed technology (CT) 30
Gas
- fluidized beds, velocity in 7
- phase olefin polymerization reactor 77

H

Horizontal pipes 24
Hydrodynamic
- particle interaction 48
- studies 67

I

Interparticle forces (IPF), effect of. 100

L

Langrangian particle tracking 13
Lattice Boltzmann method of simulation 36
Leg time method of measurement 95
Liquid feed injection 67
Long pipelines, investigation of 18

M

Mass transfer in circulating fluidized beds 52
Muliparticles, sedimentation of cylindrical 36
Multiphase opaque systems 30

O

Opaque multiphase flows 30
Ozone decomposition 52

P

Particle(s)
- clusterings in down flow reactors 48
- image velocimetry (PIV) 90
- interaction, hydrodynamic 42
- mixing 77
- measurement of velocity and direction of 95
- particle/gas flow characteristics 18
- temperature behavior of 71
- velocity 67

Pipes, horizontal, inlet regions of 24

PIV (particle image velocimetry) 90
Polymer particle mixing and segregation 77
Polyolefin synthesis 71
Pulverization of rubber granules 83

R

Reactor (s)
  down flow 42
  gas phase olefin polymerization 77
Riser, 0.45m 67
Rotating fluidized bed coater (RFBC). 112
Rubber granules, pulverization of 83

S

Saltation phenomena 24
Separators, cyclone 13
Simulation (s)
  DEM method of simulation 71
  of multiphase opaque systems 30
  of sedimentation of cylindrical particles 36
Solid state shear extrusion (SSSE). 1, 83
Suspension flows, characterization of 90

T

Three-dimensional space, multiparticles in a 36

V

Velocimetry, particle image (PIV) 90
Velocity
  continuous measurement of 95
  particle. 7, 67, 95